Climate Diplomacy at the Crossroads

Environment and Society

Series Editor
Douglas Vakoch

As scholars examine the environmental challenges facing humanity, they increasingly recognize that solutions require a focus on the human causes and consequences of these threats, and not merely a focus on the scientific and technical issues. To meet this need, the Environment and Society series explores a broad range of topics in environmental studies from the perspectives of the social sciences and humanities. Books in this series help the reader understand contemporary environmental concerns, while offering concrete steps to address these problems.

Books in this series include both monographs and edited volumes that are grounded in the realities of ecological issues identified by the natural sciences. Our authors and contributors come from disciplines including but not limited to anthropology, architecture, area studies, communication studies, economics, ethics, gender studies, geography, history, law, pedagogy, philosophy, political science, psychology, religious studies, sociology, and theology. To foster a constructive dialogue between these researchers and environmental scientists, the Environment and Society series publishes work that is relevant to those engaged in environmental studies, while also being of interest to scholars from the author's primary discipline.

Recent Titles in the Series

Climate Diplomacy at the Crossroads: A Multidisciplinary Perspective, edited by Şuay Nİlhan Açıkalın

Concrete Mirage: Governance, Equity, and Sustainable Cities in the United States Southwest, by Erin Heinz

Building Sustainable Agrifood Systems and Resilient Rural Communities in Japan, edited by Raymond Jussaume, Keiko Tanaka, and Kiyohiko Sakamoto

Strategic Rebuilding and Affordable Housing in the Muslim World, edited by Akel Ismail Kahera, Sultan Barakat and Ghadanfar

Encounters with the Posthuman and the Environment, edited by İnci Bilgin Tekin and Zeynep Talay Turner

Canals in a Changing Britain: Construction, Culture, and Environment, 1760-1968, by Jules P. Gehrke

Urban Narratives About Nature: Socio-Ecological Imaginaries Between Science and Entertainment, edited by Carlos Tabernero

The Age of Capitalism, Consumer Culture, and the Collapse of Nature in the Anthropocene, by Jack Thornburg

Sustainable Energy Development: Technology and Investment, edited by Kimarie Engerman, Elizabeth A. Keenan, Dmitry Kurochkin, Elena V. Shabliy

Making Nature Social: Towards a Relationship with Nature, by Rembrandt Zegers

Earth Polyphony: Law, Ecocriticism, and Eco-Activism in Postcolonial India, by Suhasini Vincent

Staying Together: Natureculture in a Changing World, Edited by Kaushani Mondal

Learning in the Anthropocene: Reimagining Education in the Twenty-First Century, by Carl A. Maida

Climate Diplomacy at the Crossroads

A Multidisciplinary Perspective

Edited by Şuay Nİlhan Açıkalın

BLOOMSBURY ACADEMIC
NEW YORK • LONDON • OXFORD • NEW DELHI • SYDNEY

BLOOMSBURY ACADEMIC
Bloomsbury Publishing Inc, 1359 Broadway, 12th Floor, New York, NY 10018, USA
Bloomsbury Publishing Plc, 50 Bedford Square, London, WC1B 3DP, UK
Bloomsbury Publishing Ireland, 29 Earlsfort Terrace, Dublin 2, D02 AY28, Ireland

BLOOMSBURY, BLOOMSBURY ACADEMIC and the Diana logo are trademarks of Bloomsbury Publishing Plc

First published in the United States of America 2025

Cover image:
© Malekas85/iStock

Library of Congress Control Number: 2025947316

ISBN: HB: 978-1-66697-823-0
ePDF: 979-8-21625-275-7
eBook: 978-1-97876-829-1

Typeset by Deanta Global Publishing Services, Chennai, India
Printed and bound in the United States of America

For product safety related questions contact productsafety@bloomsbury.com.

To find out more about our authors and books visit www.bloomsbury.com and sign up for our newsletters.

"To the future generations, the most precious harvest of our efforts and the living legacy of the seeds we plant today—may they inherit a world nurtured by our care, courage, and commitment to a sustainable future."

Contents

Foreword

Forewords of most of the books and works on climate diplomacy start with how climate change is important. I do believe that we have enough narrative and storytelling on how the consequences of climate change are an undeniable reality, which finds a ground in diplomacy as well. Climate change is a multidimensional phenomenon that has been the scope of the climate regime and global cooperation. Over the years, climate diplomacy has changed its face and pace, showing ups and downs when it comes to playing a determinant role for actors and going beyond the climate regime.

As a scholar of International Relations who has worked on the social consequences of climate change in recent years, I have noticed that we need to talk more from both interdisciplinary and multidisciplinary perspectives. In this respect, the idea of the book of climate diplomacy framework is to open a new window.While we were preparing this book, the United States of America once more decided to withdraw from the Paris Agreement by January 2025. In addition to this, societies are having a hot debate over whether climate change is a hoax or not. In other words, the climate regime, with its properties, is on the edge of transformation. I am personally optimistic as a believer, but it seems that the climate regime needs our personal effort more than ever. It doesn't matter whether you are a scholar or a policymaker; every individual can help tackle the consequences of climate change. Thus, it is clear that we—scholars—need to talk about climate change in various aspects.

This book aims to answer multidisciplinary questions in climate diplomacy. In this respect, readers will find three main sections: debates from theory to practice, tools and actors, governance and leadership in climate diplomacy. Distinguished authors bring innovative perspectives to the literature through our book. Personally, once again, I would like to express my gratitude to them for their efforts and patience in working on climate diplomacy. Last but not least, I want to thank my family for their endless support. Consequently, climate diplomacy is an urgent motive for actors who are based on the idea of taking action "here and now!"

Assoc. Prof. Dr. Şuay Nilhan Açıkalın
Editor
January 2026

Part I

From Theory to Practice

1

History of Climate Diplomacy

Halit Burç Aka and Sezin İba Gürsoy

1.1 Introduction

Climate change is not merely an environmental challenge; it is a profound threat to global security. By exacerbating water and food scarcity, it undermines economic stability, particularly in production sectors, and risks triggering large-scale migration. These cascading effects threaten global peace and stability, making climate change a transnational issue that no single state can address in isolation. This reality has necessitated a shift from *raison d'état* to *raison de système*, wherein states prioritize collective action and the stability of the international order over narrow national interests.

The international climate regime, as it exists today, is a product of this shift, reflecting the triumph of *raison de système* in addressing a transnational challenge that transcends the capacities of individual states. Consequently, the international community has developed mechanisms to foster cooperation among states, giving rise to climate diplomacy and specialized international institutions. The adoption of the Paris Climate Agreement in 2015 marked a significant milestone, expanding the scope of climate diplomacy to include non-state actors and further institutionalizing global efforts to combat climate change.

Today, the international climate regime stands as a testament to decades of diplomatic efforts. However, this regime did not emerge spontaneously. It is the product of complex interactions among states and, increasingly, non-state actors, who have collectively shaped the rules, norms, and values governing climate affairs. Understanding the historical evolution of climate diplomacy is therefore essential for analyzing how international norms and institutions merge, evolve, and endure beyond the immediate interests of individual states.

This chapter aims to trace the historical development of climate diplomacy, highlighting key milestones and turning points. By situating these developments within a historical context, this chapter contributes to a deeper understanding of how international cooperation on climate change has evolved and what challenges lie ahead.

On January 20, 2025, President Donald Trump's executive order to withdraw the United States from the Paris Agreement sparked significant concerns about the future of climate diplomacy. While Trump had signaled his intention to withdraw from the agreement during his first term, the decision to formalize this stance through legal mechanisms marked a pivotal moment in American climate policy. However, attributing this move solely to Trump's leadership would be an oversimplification. For instance, Falkner (2005) had already noted two decades earlier that the United States had distanced itself from its role and responsibility as a global environmental leader. Furthermore, the George W. Bush administration (2001–9) also adopted an anti-climate regime stance (Buzan & Falkner, 2022, p. 28). The key difference between Bush and Trump lies in the latter's decision to formalize this stance through legal mechanisms. MacNeil and Paterson (2020) describe American engagement in climate diplomacy as oscillating between hegemonic multilateralism, where the United States asserts its preferences on the global stage, and an isolationist stance of outright rejection. In fact, this oscillation reflects nothing more than two competing dynamics: *raison d'état* versus *raison de système*.

Certainly, given that global constitutive and regulatory institutions have been established to address climate change, President Trump's decision will be regarded as a turning point in the history of climate diplomacy. However, some scholars dispute this interpretation. For instance, Laura Von Allwörden (2024), in her empirical study, argues that Trump's withdrawal could reinforce support for climate-related norms and institutions. She contends that contestation of the climate regime does not necessarily weaken it; paradoxically, it may further legitimize it, as members of the global community reaffirm their commitment to the regime and actively defend it (Allwörden, 2024).

On the other hand, Hedley Bull (2022) argues that great power management is essential to the functioning of international society, ensuring order, stability, and the legitimacy of international institutions. Nevertheless, its effectiveness depends on the willingness of great powers to act collectively rather than unilaterally. Therefore, one may conclude that the United States' withdrawal from the Paris Climate Agreement could undermine the international community's capacity to create, develop, and implement rules and norms, as Trump's decision signals the United States' reluctance to uphold its international responsibilities in climate governance while also reinforcing opposition to the institutions and legal framework of climate diplomacy.

Nevertheless, the broader implications of this withdrawal remain contested. While some view it as a setback for climate diplomacy, others argue that it has inadvertently strengthened the international climate regime by highlighting the necessity of collective action and the limitations of unilateralism. This dynamic underscores the complex interplay between *raison d'état* and *raison de système* in shaping the international climate regime. This chapter addresses the following research questions: How has (1) the historical evolution of climate diplomacy shaped the international climate regime, and (2) the tension between *raison d'état* and *raison de système* shaped the historical evolution of the international climate regime? To answer this question, the chapter employs an English School framework.

1.2 English School Framework

Under this section, the concepts of diplomacy, international society, world society, international institutions, and *raison de système* are framed within the English School. In this way, the history of climate change can be understood through the lens of English School concepts.

From a traditional perspective, diplomacy is "the conduct of relations between sovereign states through the medium of officials based at home or abroad, the latter being either members of their state's diplomatic service or temporary diplomats" (Berridge & James, 2003, p. 69–70). From a broader perspective, diplomacy can be defined as the primary mechanism through which states interact, facilitating sustained and intricate relationships (Berridge & James, 2003, p. 70).

The English School recognizes war, the balance of power, international law, great power management, and diplomacy as primary international institutions. As Hurrell (2007, p. 37) states that

> Diplomacy is essential to international society because it underpins the minimal conditions and prerequisites of any cooperative social order: the capacity

> to communicate, the necessity of shared conventions for communication (linguistic and procedural), and the provision of an institutional framework to allow political negotiation to take place in strained and often very difficult circumstances. It also plays an expressive and symbolic role, especially in terms of upholding the idea of a shared international community, again particularly at times (e.g., during the Cold War or the so-called "long war against global terrorism") when its very existence is under threat and held in doubt.

Diplomacy has evolved into a sphere of activity that now includes non-state actors. In fact, such diplomatic engagements are increasingly considered within the framework of global governance. As Hurrel states, the significant increase in functional intergovernmental organizations (IGOs) and international conferences has not only introduced new actors and communication networks but has also contributed to the emergence of a global governance structure that operates independently alongside interstate diplomacy (Hurrell 2007, p. 206–2010).

While addressing crises and troubleshooting challenges are integral aspects of diplomatic practice, diplomacy is equally about initiative-taking and strategic thinking. It involves anticipating global challenges, crafting innovative solutions, and fostering international cooperation to prevent conflicts before they escalate. In that sense, climate diplomacy is "the practice and process of creating the international climate change regime and ensuring its effective operation. The evolution of climate diplomacy therefore precedes and shapes the construction of the climate regime" (Mabey et al., 2013, p. 23).

Climate diplomacy is a matter of both the international society's climate regime and the world society's global governance. At this stage, it is necessary to discuss the types of society within the English School. There is no clear consensus in the English School literature regarding the classification of societies. However, as authors, we recognize Hedley Bull's (2002) framework and foundational principles for international society as still valid. At the same time, we find Barry Buzan's (2004) distinction between international society and world society to be a more precise and explanatory approach.

Bull (2022, p. 64) argues that there exists a complex of rules that constitute the fundamental or constitutional normative principles of world politics. The first-level rule complex establishes sovereignty and territoriality as given and dominant normative principles. Since these principles are indispensable for the members of a society, they do not need to be explicitly mentioned among primary institutions.

A second-level rule complex defines the minimum behavioral conditions for states, setting the rules that regulate their coexistence. Bull's five primary institutions, along with the two implicit ones, are closely connected to the three key elements in his functional perspective on society: controlling violence (through war, balance of power, and great power management), defining property rights (via sovereignty and territoriality), and upholding the sanctity of agreements (through diplomacy and international law) (Buzan, 2014, p. 99).

Bull also discusses a third-level rule complex, which governs broader cooperation beyond mere coexistence.

> Rules of this kind prescribe behavior that is appropriate not to the elementary or primary goals of in international life, but rather to those more advanced or secondary goals that are a feature of an international society in which a consensus has been reached about a wider range of objectives than mere coexistence. Nevertheless, these rules may be said to play a role in relation to international order, inasmuch as the development of cooperation and consensus among states about these wider goals may be expected to strengthen the framework of coexistence. (Bull, 2022, p. 67–68)

Within this framework, one can find a variety of systems and agreements, ranging from the United Nations and arms control treaties to regimes and institutions dedicated to managing trade, finance, environmental issues, and numerous technical matters such as postal services, the allocation of orbital slots, and broadcast frequencies (Buzan, 2014, p. 98). These institutions are classified as secondary institutions.

Environmental concerns, including climate change, fall within this third-level rule complex. Consequently, institutions established to address environmental issues are classified as secondary institutions. Barry Buzan (2004, p. 90–139) replaces this rule-based classification with a distinction between pluralist and solidarist international societies and categorizes environmental issues within the solidarist international society framework. Although Bull and Buzan differ in their conceptual framing, they converge in their fundamental principles regarding international society.

Climate change is a major source of tension within international society. This tension arises from the difficulty of aligning states' foundational normative principles with their secondary objectives, such as international cooperation on climate change. The United States exemplifies this conflict, where national sovereignty and economic priorities frequently clash with global climate commitments. In other words, the conflict between *raison d'état* and *raison de système* is particularly evident in environmental policies.

The concept of raison de système holds a more prominent position within the English School compared to other theoretical traditions. Adam Watson extensively developed this concept: "Raison de système, the view that the rules and institutions of an international society promote the interests of a state (or of all states), developed out of raison d'état" (Watson, 1997, p. 149). In short, *raison de système* can be defined as the rationale of the international order.

Raison de système does not eliminate conflicts of interest; rather, it acknowledges that all parties benefit from addressing such conflicts within the established system, adhering to the rules and standards of conduct that govern their redefined international society (Watson, 1992, p. 240). This concept highlights the importance of systemic management, where great powers recognize that their long-term security and stability depend on maintaining a functional international order rather than pursuing unilateral gains.

While states may disagree on specific policies or strategies, they share a collective interest in preventing disorder and ensuring systemic resilience. This requires adherence to fundamental principles that uphold diplomatic engagement and institutionalized cooperation. "The concerting powers need not agree all the time; but they must all recognize the advantage of managing the pressures of the system and making the society work: they must have a sense of *raison de système*. They must be willing to implement the principles of prudence and moral obligation that have held together the society of states" (Watson, 1992, p. 322).

Historical and contemporary cases illustrate how *raison de système* operates in international society. One of the earliest manifestations of this principle was the Concert of Europe, established after the Congress of Vienna (1815). In the aftermath of the Napoleonic Wars, the major European powers—Britain, Austria, Russia, Prussia, and later France—recognized that ensuring stability across the continent required cooperation, even when their national interests diverged. Instead of pursuing outright domination over one another, they engaged in regular diplomatic consultations and agreed upon shared norms to manage conflicts and prevent large-scale wars. This system was not simply about preserving the status quo; rather, it aimed to guide controlled change while sustaining systemic stability. Even during crises—such as the Crimean War (1853–6)—the great powers prioritized the integrity of the broader order over short-term national gains, preventing the escalation of conflicts into full-scale continental wars.

Beyond security, *raison de système* also applies to economic governance, where leading states assume responsibility for maintaining financial stability. As Watson (1992, p. 304) notes: "The governments of the leading states agree that they have to guide and nurse orderly economic development throughout the world in proportion to their capacity: in other words, to accept the responsibilities of economic raison de système." A defining case of economic raison de système was the establishment of the Bretton Woods system following the Second World War. Major economic powers, particularly the United States and the United Kingdom, acknowledged that unregulated economic competition had contributed to the Great Depression and global instability. In response, they created institutions such as the International Monetary Fund (IMF) and the World Bank to coordinate monetary policies, provide financial assistance to weaker economies, and mitigate economic crises. A more recent application can be seen in the G-20 Summit, where the world's largest economies coordinate policies to safeguard global financial stability. Following the 2008 financial crisis, G-20 leaders implemented regulatory measures, stimulus programs, and debt relief initiatives to prevent economic collapse, demonstrating a shared commitment to systemic economic stability rather than allowing financial disorder to escalate into deeper crises.

Similarly, raison de système extends to security dynamics, particularly in the responsibility of great powers toward weaker states. As Watson (2007, p. 68) argues, "the great powers or superpowers in an international society are responsible for the security of the weak as part of raison de système." A contemporary illustration of this is NATO's response to Russia's invasion of Ukraine in 2022. While individual member states—notably the United States, the United Kingdom, Germany, and France—have their own strategic interests, their collective actions align with raison de système in ensuring European security and systemic stability. By providing military, economic, and humanitarian aid, NATO countries seek to prevent the erosion of sovereignty and territorial integrity, thereby reinforcing the principles that underpin the international order. Their involvement is not solely about supporting Ukraine but about maintaining stability in the broader European security framework, ensuring that disorder does not spread beyond the conflict zone.

Through these cases, raison de système emerges as a guiding principle in both political and economic governance. Whether in diplomatic cooperation among great powers, economic coordination, or security commitments, this concept underscores the need for leading states to manage systemic pressures rather than allowing crises to spiral out of control.

The fact that environmental problems know no borders has led scientists, politicians, civil society, and diplomats to international cooperation. The climate crisis is one of the greatest environmental challenges of the twenty-first century. This is where climate diplomacy comes in its new form. Climate diplomacy involves utilizing diplomatic efforts and strategies to tackle global climate change and its effects on international relations. It encompasses the negotiation and enforcement of climate-related policies, agreements, and treaties at the global level, as well as fostering cooperation between nations and other stakeholders to mitigate greenhouse gas emissions, adapt to climate impacts, and support sustainable development. In the past ten years, it has been observed that climate diplomacy has evolved into a more comprehensive strategy. This first chapter of the book aims to set out contemporary definitions of climate diplomacy, recognizing that climate change is a wide-ranging issue involving engagement at multiple levels. This chapter is also important in terms of revealing the application areas of climate diplomacy and the problems that can be experienced in these areas.

1.3 Development of Climate Diplomacy

During the twentieth century, industrialization and technological advances significantly disrupted the Earth's ecological balance, leading to an unprecedented increase in human impact on the environment. In addition, population growth, pollution, and urbanization have exacerbated environmental challenges such as water scarcity, biodiversity loss, and climate change, making them persistent global problems. The transboundary nature of these environmental issues, which affect all aspects of human life, has necessitated international cooperation among scientists, policymakers, civil society, and diplomats. It has become clear that challenges such as pollution, loss of biodiversity, depletion of natural resources, and climate change are not confined to a single nation but are global in scope. Addressing these issues effectively requires not only the efforts of NGOs and individuals but also coordinated action by international organizations and the alignment of national policies.

The increase in the severity of the problems and the changes in the international system have had an impact on the emergence of environmental issues as a subject for diplomacy. The classical image of diplomacy has been replaced by an image of hybrid diplomacy with a multiplicity of actors, a multiplicity of issues, and a multiplicity of practices. The transition from an international to a world society in the context of climate diplomacy is marked by a shift from state-centric negotiations to more inclusive, multilateral approaches involving a wider range of actors, including non-state entities and civil society. This shift is evident in the evolution of climate agreements and the increasing role of global cooperation in addressing the challenges of climate change.

The threats may not necessarily cooperate, but at least they are serious and persistent enough to shake key actors out of their traditional slumber (Weiss & Burke, 2011, p. 1065). Climate change is a pressing threat that transcends national boundaries and requires international cooperation and coordinated solutions at all levels. Climate change, as with other environmental problems, is essentially a matter of multilateral politics. Foresighted approaches to solving international problems and crisis management are considered an important part of diplomatic practice. Climate diplomacy refers to international efforts and negotiations aimed at combating climate change by strengthening cooperation between nations. It involves a complex interplay of scientific knowledge, political interests, and global governance structures through the process of building the international climate regime and making it work effectively. Over the past few decades, scientific understanding of the climate problem has improved radically, and since the late 1970s, there have been ongoing diplomatic discussions and numerous formal agreements on the issue. Since the 1970s, climate policy has been diplomatically replaced by scientific awareness and socioeconomic factors. Over the last few years, several meetings and conferences have been organized to deal with the climate issue. As an independent body, the Intergovernmental Panel on Climate Change (IPCC) bridges the gap between science and diplomacy. However, despite scientific agreement on the risks of climate change, diplomatic initiatives are often limited by national interests that can prevent diplomats from reconciling their differences. In this sense, the international community's response to climate and environmental issues began with the UN Environmental Declaration in 1972 and reached a significant level with the 2015 Paris Climate Agreement. The development of climate diplomacy, therefore, precedes and shapes the construction of the climate regime. (Mabey & Born, 2013, p. 23)

1.4 Key Milestones of Climate Diplomacy

International diplomacy on climate change began more than fourty years ago, reflecting the scientific, economic, and political complexity of the issue. This diplomacy can be traced through key milestones and events that have been in place since the issue's rise to international prominence. Climate diplomacy is the use of diplomatic tools to address climate challenges. This includes negotiating, building alliances, and encouraging partnerships to collaborate. The international community initially sought to address these challenges with goals such as mandatory emission reductions and the establishment of an international climate regime. As a result of successful international negotiations on ozone depletion, climate change, and other issues, work began in the early 1990s to establish the UNFCCC. Based on this universal agreement, the aim was to establish a set of regulatory instruments by introducing increasingly stringent rules. This section covers the United Nations Framework Convention on Climate Change (UNFCCC), the Kyoto Protocol, and the Paris Agreement, as well as other key global climate change events such as world summits.

Environmental policy has been the subject of international diplomacy since the 1960s, when global environmental degradation began to attract the attention of industrialized countries. Environmental issues first entered international diplomacy within the framework of the United Nations (UN) and its various specialized agencies. Following growing concerns about the environment, the Stockholm Summit, the first multilateral summit to address environmental issues, was organized in 1972. Setting the stage for future climate diplomacy efforts, it was one of the first major international meetings to address environmental issues. One of the major achievements of the Stockholm Conference was the creation of the United Nations Environment Programme (UNEP), which has, over time, contributed to the progress of subsequent world climate conferences.

1.5 Stockholm Conference and Rio Summit

The early 1990s saw an increase in environmental initiatives. The UN Declaration on the Human Environment and the Brundtland Report contributed to the UN Conference on Environment and Development (UNCED) held in Rio in 1992. The conference put issues such as global environmental change and degradation on the world agenda. This conference, known as the Earth Summit, led to the creation of the UNFCCC, which established a framework for international climate negotiations. The 1992 Convention was signed to "stabilize greenhouse gas concentrations in the atmosphere at a level that would prevent dangerous human interference with the climate system," entered into force on March 21, 1994. It requires signatories to limit their greenhouse gas (GHG) emissions to the level that can be absorbed by the earth's biosphere and calls on developed countries to take the lead because of their historical responsibility for their disproportionate emissions. The Convention is a foundational document that outlines two closely related policy approaches to tackling global warming: mitigation and adaptation. Mitigation focuses on tackling the root causes of climate change, while adaptation aims to deal with its inevitable impacts. The Convention urges ratifying nations to convene regularly through the Conference of the Parties (COP). Decisions are reached through negotiations within the COP, which gathers annually to assess the progress of UNFCCC implementation.

Effective representation in the UNFCCC process is the first stage of climate diplomacy (Mabey & Born, 2013, p. 24). The other stage is the "common but differentiated responsibilities" (CBDR) notion, which is one of the main principles of the Convention. It is very important in negotiations and is also known as "differentiation." Developing countries have interpreted CBDR as the historical responsibility of states to address the climate problem. Moreover, developed countries have tended to resist the concept of historical responsibility and the rigid North-South divide, focusing instead on capacity and current and future contributions to climate change (Brunnée & Streck, 2013, p. 590). Historically, the most difficult aspect of international discussions on climate change has been the issue of equity. The general principle is that industrialized countries bear the primary responsibility for the climate problem because of their far greater historical aggregate per capita emissions and their declining share of the world's population. The concept of faster adaptation for developed countries implies that poorer countries have some right to "common but differentiated exceptions." When it comes to international equity, an important underlying concern is the impact of wealth and poverty on different levels of activity that affect equilibrium greenhouse gas emissions. Unsatisfactory agreement on how to apply the principle makes climate change a difficult issue in international diplomacy. As the rate of growth of energy emissions in developing countries has been and continues to be high, the most recent data suggest that developing countries together are now responsible for most energy emissions.

1.6 Kyoto Protocol Pluralist International Society

Five years after the Summit, in COP3, the "Kyoto Protocol" was adopted to extend the Framework Convention on Climate Change. This agreement required signatories to reduce emissions of greenhouse gases that cause global warming by varying amounts by 2012. It was a major milestone in international climate action, establishing mechanisms such as carbon trading to support emissions reductions (The United Nations, 1998). The protocol focused on developed countries like the United States and their carbon emissions, rather than placing the greater burden on developing countries like China or India. However, the agreement, which imposed no limits on countries that continued to develop, was deeply wounded when the United States failed to ratify the protocol.

The Kyoto Protocol is the most controversial environmental agreement in world history. According to the principles and provisions of the Convention, developed countries are largely responsible for the current high levels of greenhouse gas emissions in the atmosphere. Under the principle of CBDR, it places a heavier burden on these countries. Developing countries have insisted on its continuation because it contains emission reduction commitments for developed countries listed beyond the first commitment period 2008–12, i.e., beyond the Kyoto (Brunnée & Streck, 2013, p. 590). To put it mildly, the consequences of Kyoto are mixed. The Protocol has failed to reduce greenhouse gas emissions to the extent required. This is because the ambitious targets of the Kyoto Protocol apply only in the short term (2008–12). Moreover, one of the main shortcomings of the Protocol is the lack of compliance incentives and enforcement mechanisms. It is thought that the Protocol imposes relatively high costs and will only create short-term benefits. (Olmstead & Stavins 2006, p. 35). In addition, Kyoto has deepened the CBDR doctrine formulated in the UNFCCC. In-line with the CBDR doctrine, Kyoto rigidly divided countries into two groups. This reinforced the existing ideological North-South divide by creating a so-called firewall. Olmstead and Stavins argue that Kyoto's distinction between richer and poorer countries is clearly outdated and inaccurate.

Since the mid-2000s, the scope of climate diplomacy has broadened and deepened again, driven by the need to respond to the increasing impacts of climate change and to ensure the growth of a low-carbon economy (Mabey & Born, 2013, p. 27). The most significant international environmental development in 2000 was the UN Millennium Declaration, which strengthened commitments to protect the natural environment and other global issues. Climate initiatives, which have gained momentum especially since the early 2000s, tend to internationalize. COP12 in Montreal recognized that Kyoto alone was not enough to tackle global warming, and COP13 in Bali established the Bali Action Plan (BAP) as the basis for an innovative two-way climate negotiation process. It distinguished between these two categories of countries, calling for mitigation commitments from the former and mitigation actions from the latter (Rajamani, 2016). The main objective of the roadmap is to reach a legally binding, comprehensive climate agreement in Denmark in 2009 that would replace the Kyoto Protocol after 2012 and ideally include all major greenhouse gas emitters. BAP emphasized the need for a "shared vision" for global climate action, ensuring that it is consistent with the provisions and principles of the Convention, particularly the principle of CBDR. Notably, it is the first international climate decision in which the United States participated, despite its previous strong opposition. This has raised hopes that the United States could re-engage in the post-2012 climate negotiations. Another step on the road to Paris was the 2009 Copenhagen Conference (COP-15), which was the lowest point in the history of the climate regime. As Savaresi noted, trust in the Copenhagen conference was broken, and the process was poorly managed diplomatically (Savaresi, 2016). According to Bäckstrand and Lövbrand (2019, p. 7), the Copenhagen Accord's focus on "nationally appropriate" mitigation commitments represents a significant break with the KP's multilateral coordination of national carbon budgets, emission targets, and mitigation techniques. COP15 removed the Kyoto Protocol's firewall between Annex I and non-Annex I countries, paving the way for a broader mitigation agreement and setting the stage for the Paris Agreement (Falkner, 2016).

1.7 Paris Agreement-2015 Birth of World Society

Another milestone is the Paris Agreement, signed during COP21, which was adopted by 196 Parties on December 12, 2015; it officially came into effect in 2016. This landmark agreement is a major success of multilateral climate diplomacy, bringing together almost all countries to tackle climate change together. This agreement outlines how nations will fulfill their commitments under the UNFCCC beyond 2020. The goals of the Paris Agreement are to stabilize global warming above pre-industrial levels at well below 2 degrees Celsius, while pursuing efforts to limit the temperature increase to 1.5 degrees Celsius, recognizing that this would significantly reduce the risks and impacts of climate change, including the vulnerability of natural and human systems and the ability to adapt to warming. Unlike the Kyoto Protocol, the Paris Agreement allows countries to set their own climate targets through nationally determined contributions (NDCs), which are regularly reviewed and updated. The Paris Agreement is an international treaty on climate change that allows national governments to meet their commitments in their own countries by adopting a more flexible approach. The Paris Agreement strengthened key measures to enable countries to meet their commitments. These strengthened measures were discussed and implemented in the detailed Katowice Rulebook. The agreement paves the way for global collective action that will require significant international policy coordination to effectively address climate change by forcing high-emitting countries in the North and South to reduce their emissions (Bäckstrand & Lövbrand, 2019, p. 8).

1.8 From Kyoto to Paris: From International Community to World Community

The large number of new studies on global climate change issues, the increase in the number and scope of transnational networks, the growth of entrepreneurial private authority, various initiatives in transnational climate change governance, and a wide range of activities are crucial for creating new opportunities for climate cooperation (Koehone & Victor, 2016, p. 574). In the international community, climate negotiations have been driven primarily by national interests, sovereignty concerns, and power dynamics between states, whereas the world community has evolved into a more inclusive and collaborative approach. The concept of the world community captures the role of non-state actors, including international organizations, NGOs, businesses, and civil society, in global climate governance. From the Kyoto Protocol to the Paris Agreement, climate governance has been seen as going beyond traditional diplomacy to include a wide range of stakeholders. In Paris, a multifactorial and multilayered perspective was adopted, with state commitments, sub-national actors, multinational corporations, and transnational networks.

Looking at the differences between Kyoto and Paris, one can see the contradiction between the binding, state-centered approach of Kyoto and the flexible, global community standards of Paris. The 1997 Kyoto Protocol set relatively modest targets for developed countries and exempted developing countries from similar obligations. Some countries (e.g., Canada) also failed to meet their Kyoto targets, while Russia and other post-Soviet states reduced their emissions without any effort. In contrast to Kyoto, the Paris Agreement involves all countries in the effort to reduce emissions, while it still retains a degree of distinctiveness. The Paris Agreement also states that "climate justice is difficult and controversial." The UNFCCC and the Kyoto Protocol made the CBDR principle work by requiring only developed countries to take on ambitious greenhouse gas reduction targets. This kind of differentiation has proved quite controversial over the years. The Paris Agreement takes the differentiation a step further by adapting it to the characteristics of each of the Durban pillars: mitigation, adaptation, finance, technology, capacity building, and transparency (Rajamani, 2016). The term climate justice has been used to refer to assessments of distributive and corrective justice associated with both the impacts of climate change and the policies associated with responding to climate change. Climate justice is about how to share the burdens associated with a global transition to low-carbon societies (Savaresi, 2016).

The Paris Agreement represents the culmination of decades of global efforts to address climate change. It is seen as a challenge for action that requires an effective solution to a major problem of international policy coordination. According to Falkner (2016), more than two decades after the adoption of the convention, an effective climate agreement is not about finding quick solutions to the emissions problem, but rather about setting the structure for a long-term technological and economic transformation. This agreement outlines how nations will meet their commitments under the UNFCCC beyond 2020. According to Hale (2020), the Paris Agreement brought the logic of cooperation to the fore and raised the possibility that similar catalytic institutions could facilitate cooperation in other areas of global governance characterized by similar problem structures.

The Paris Agreement takes a diplomatic approach to risks that countries cannot address by acting alone. While the agreement does not "solve" climate change, it was a critical turning point in climate diplomacy. The challenge now is to deliver on the promises made in Paris. In this context, the agreement promised a more realistic path to globally coordinated emissions reductions and succeeded in bringing international climate policy more in-line with reality. Success now largely depends on what happens at the national and regional levels. The challenge is to close the gap between developed and developing countries.

1.9 COP26 and COP27: Strengthening the World Community

Due to Covid-19, COP26 was postponed and was held in Glasgow from October 31 to November 13, 2021. Some believe that this postponement will give countries time to develop more ambitious targets and accelerate low-carbon actions to ensure a green and resilient recovery from Covid-19 (Tskipurishvili, 2022, p. 2). Two weeks of intense negotiations focused on mitigation, adaptation, loss and damage, finance, and cooperation. COP26 concluded with all countries adopting the Glasgow Climate Pact to keep 1.5 degrees Celsius alive and complete the missing elements of the Paris Agreement. The Glasgow Climate Pact, combined with increased ambition and action by countries, means that keeping 1.5 degrees Celsius in sight and scaling up action to tackle climate impacts will only be achieved through concerted and urgent global efforts. In November 2022, Egypt hosted the 27th Conference of the Parties (COP27) to the UNFCCC in Sharm el-Sheikh. COP27 resulted in the recognition of the seriousness of the global climate challenge and the value of multilateral action to address this global threat. Egypt's COP27 presidency brought the practical feasibility of comprehensive action closer, moving from diplomacy and planning to implementation.

At COP26, world leaders and governments renewed their commitment to global action on adaptation. At COP27, they sought to demonstrate their willingness to invest in making the fundamental progress needed and to engage all parties in building resilience and helping the most vulnerable communities (Locci & Jammaoul, 2023, p. 120). The decisions taken at COP26 and COP27 strengthen the norms of justice and responsibility of the global community. The Just Transition Charter adopted at COP26 aims to ensure that no one is left behind in the transition to net-zero emissions economies. At COP27, it was decided to establish a Loss and Damage Fund for vulnerable countries most affected by climate disasters. This fund aims to help developing countries cope with the negative impacts of climate change and is seen as an important step toward climate justice. These initiatives reflect the efforts of the global community to strengthen the principles of equity and responsibility in the fight against climate change.

1.10 Climate Diplomacy in Transition: Oscillating Between *Raison d'etat* and *Raison de Système*

The raison de système is expected to prevail over the raison d'état in the international climate regime due to three key factors: the transnational nature of climate change, which makes it insurmountable through unilateral action, even by great powers; the establishment of deep and robust cooperation within international society (raison de système); and the historical process through which climate change has evolved into a fluid and dynamic space between international society and global society. Together, these factors limit the influence of raison d'état and reinforce the primacy of raison de système in addressing global climate challenges. Be that as it may, historical evidence demonstrates that *raison d'état* has been more decisive in climate diplomacy.

At first glance, one might expect *raison de système* to dominate climate diplomacy, as it is based on the rational consent of states to cooperate within an established international framework. After all, the primary objective of climate diplomacy is to establish institutions that address climate change and its associated challenges. Is there any state immune to the adverse effects of climate change? Certainly not. However, it is equally unsurprising that states prioritize their short- and medium-term national interests over long-term environmental concerns. As a result, economic interests tend to outweigh environmental considerations in climate diplomacy.

Another key takeaway from the history of climate diplomacy is that its acquis has been shaped and developed through the conflict between *raison d'état* and *raison de système.* More softly put, the history of climate diplomacy has essentially evolved through the competition between national interests and the interests of international society. A clear example of this is the 1992 UNFCCC, which was adopted at the Rio Earth Summit, laying the legal foundation for the climate regime. For the first time, international institutions actively positioned themselves to address environmental challenges, which were recognized as part of the new post-Cold War threats. The convention required states to make general commitments to reduce emissions. However, despite agreeing to cut greenhouse gas emissions, developed economies hesitated to include binding commitments in the treaty. Some states—particularly the United States—resisted more concrete obligations due to concerns over economic losses. In contrast, developing economies successfully introduced the principle of "Common but Differentiated Responsibilities" (CBDR).

Another example is the 2009 Copenhagen Accord. Despite high expectations at COP-15, the summit resulted in a non-legally binding agreement. The United States opposed binding targets, prioritizing its economic and political interests. Similarly, China and India were reluctant to accept binding targets, arguing that development should take precedence. These countries demanded more flexible regulations rather than strict carbon reduction commitments. Meanwhile, less developed economies took a strong stance on climate justice, calling for greater financial and technological support. As these two cases illustrate, climate diplomacy continues to evolve through the ongoing competition between *raison d'état* and *raison de système.*

The history of climate diplomacy demonstrates that combating climate change has become one of the broader objectives of international society. Through diplomacy and great power management—both recognized as primary institutions of international society—norms, rules, and values in the field of climate governance have been continuously developed. Key examples include the CBDR principle, climate justice, and greenhouse gas reduction mechanisms.

Additionally, climate negotiations are a significant example of multilateral diplomacy and norm-building processes, as emphasized by the English School. The Paris Climate Agreement and COP summits exemplify the process in which states assume shared responsibilities and contribute to norm formation. COP summits can be seen as platforms where international society convenes to establish common norms. Decisions such as the Loss and Damage Fund demonstrate that international society is beginning to take on the responsibility of protecting weaker states.

Meanwhile, the Paris Climate Agreement has marked a noteworthy moment where members of the world society, albeit in a limited manner, have contributed to the formation of institutions such as norms, rules, and principles in climate governance. Despite its shortcomings, one can speak of a climate regime, though it is still premature to claim the existence of a fully functional system of global climate governance. In examining climate diplomacy, we recognize that since the Paris Climate Agreement, a limited interaction has emerged between international society and world society, particularly in environmental matters such as climate change. However, there is not yet sufficient observation to determine whether this interaction is complementary or conflictual.

Our assessment that the international community has, in one way or another, achieved certain gains that are contested by some authors. For instance, according to Andrew Hurrell (2007), states and the system of sovereign states are increasingly failing to manage an interconnected world, and particularly the global environment, in an effective and equitable manner. The international community has been unable to establish a suitable political framework for the collective governance of environmental issues. Moreover, states are not successful in addressing environmental challenges on their own. While Hurrell acknowledges the Kyoto Protocol as a significant step in the fight against climate change, he also notes its limited effectiveness. However, he concedes that such agreements provide a political framework for more effective cooperation. Ultimately, he emphasizes that the state and state interests remain central in addressing climate change.

Bibliographies

Bäckstrand, K., & Lövbrand, E. (2019). The road to Paris: Contending climate governance discourses in the post-Kyoto era. *Journal of Environmental Policy & Planning, 21*(5), 519–532.

Berridge G. R., & James, A. (2003). *A dictionary of diplomacy*. Palgrave Macmillan.

Brunnée, J., & Streck, C. (2013). The UNFCCC as a negotiation forum: Towards greater effectiveness? *Climate Policy, 13*(3), 837–857.

Bull, H. (2002). *The anarchical society: A study of order in world politics* (3rd ed.). Palgrave.

Buzan, B. (2004). *From international to world society? English School theory and the social structure of globalisation*. Cambridge University Press.

Buzan, B. (2014). *An introduction to the English School of International Relations: The societal approach*. Polity Press.

Buzan, B., & Falkner, R. (2022). Great powers and environmental responsibilities: A conceptual framework. In R. Falkner & B. Buzan (Eds.), *Great powers, climate change, and global environmental responsibilities* (pp. 14–48). Oxford University Press.

Falkner, R. (2005). American hegemony and the global environment. *International Studies Review, 7*(4), 585–599.

Falkner, R. (2016). The Paris Agreement and the new logic of international climate politics. *International Affairs, 92*(5), 1107–1125.

Hale, T. (2020). Catalytic cooperation: The Paris Agreement in comparative perspective. *International Studies Quarterly, 64*(3), 550–563.

Hurrell, A. (2007). *On global order: Power, values, and the constitution of international society*. Oxford University Press.

Keohane, R. O., & Victor, D. G. (2016). Cooperation and discord in global climate policy. *Nature Climate Change, 6*(6), 570–575.

Locci, E., & Jammoul, H. (2023). *Climate diplomacy: International and national responses*. Dar Abaad.

Mabey, N., Gallagher, L., & Born, C. (2013). The evolution of climate diplomacy and the international climate regime. In *Understanding climate diplomacy: Building diplomatic capacity and systems to avoid dangerous climate change* (pp. 23–27). E3G. http://www.jstor.org/stable/resrep17706.6

MacNeil, R., & Paterson, M. (2020). Trump, US climate politics, and the evolving pattern of global climate governance. *Global Change, Peace & Security, 32*(1), 1–18.

Olmstead, S. M., & Stavins, R. N. (2006). An international policy architecture for the post-Kyoto era. *American Economic Review, 96*(2), 35–38.

Rajamani, L. (2016). Ambition and differentiation in the 2015 Paris Agreement: Interpretative possibilities and underlying politics. *International & Comparative Law Quarterly, 65*(2), 493–514.

Savaresi, A. (2016). The Paris Agreement: A new beginning? *Journal of Energy & Natural Resources Law*, *34*(1), 16–26.

Tskipurishvili, A. (2022). *New climate diplomacy under the Paris Climate Agreement.* Geneva Global Policy Brief No. 1. https://ceje.ch/files/8616/4450/0110/University_of_Geneva_-_GGPB_N1-2022_-_A._Tskipurishvili.pdf

United Nations. (1998). *Kyoto Protocol to the United Nations Framework Convention on Climate Change.* New York: United Nations. https://unfccc.int/resource/docs/convkp/kpeng.pdf

von Allwörden, L. (2024). When contestation legitimizes: The norm of climate change action and the US contesting the Paris Agreement. *International Relations*, *39*(1), 52–75. Advance online publication. https://doi.org/10.1177/00471178231222874

Watson, A. (1992). *The evolution of international society*. Routledge.

Watson, A. (1997). *The limit of independence: Relations between states in the modern world*. Routledge.

Watson, A. (2007). *Hegemony and history*. Routledge.

Weiss, T. G., & Burke, M. J. (2011). Legitimacy, Identity and Climate Change: moving from international to world society? *Third World Quarterly*, 32(6), 1057–1072. https://doi.org/10.1080/01436597.2011.584721

2

Traversing Through India's Climate Diplomacy

An Analysis from a Human Rights Perspective

Amit Upadhyay and Abhinav Mehrotra

2.1 Introduction: The "Indian" Approach to Climate Diplomacy

Climate change is proving to be one of the biggest challenges currently faced by nations, with climate, energy, and development issues becoming core foreign policy concerns for nations as they realize the urgent need for domestic climate action and international cooperation to mitigate the adverse effects of climate change. The Covid-19 pandemic led to the global economy shrinking by 5.2 percent, pushing millions into poverty (World Bank Group, 2020). The increase in extreme climate events such as forest fires, extreme heatwaves, floods, droughts, and locust attacks in many parts of the world shows the magnitude of the climate crisis and its transboundary impacts. It shows the importance of international cooperation, as individual country-led climate action may not be sufficient to address climate change. Climate diplomacy is becoming increasingly complex because certain nations see it as an environmental issue, while others view it as an issue deeply rooted in justice and equity, as well as an economic and technological challenge. Consequently, climate action is fraught with differences in interests and continues to impact countries, their politics, and policies in their efforts to negotiate climate partnerships with other countries. In the last few decades, India has significantly diversified and expanded its climate diplomacy to become an indispensable force in global climate politics (Dubash, 2019).

India has become an active player in global climate politics. It is not only raising its climate interests internationally but also raising concerns of other developing countries whose interests India seeks to represent at the international level. It has remained consistent throughout its engagement with global environmental issues. Its consistent position reflects a commitment to principles that address the historical and structural inequities embedded in international environmental governance frameworks (Gupta et al., 2015). However, India's firm positions are often perceived as obstacles to achieving consensus within the international community. Still, India, as a member state, has justified its foreign policies with the broader philosophy of justice and equity that prioritizes the needs and developmental aspirations of the Global South. Such a position is also reflected in India's National Environmental Policy of 2006 (Singh and Mishra, 2015), which outlines its strategic approach to addressing environmental challenges. This policy emphasizes five core elements of India's approach to climate change issues, including adherence to the principle of Common but differentiated responsibilities (CBDR) and respective capabilities, as enshrined in international agreements; prioritization of the Right to Development, particularly for developing nations; advocacy for equal per capita entitlements to global environmental resources; reliance on multilateral approaches to foster inclusive and cooperative decision-making; and engagement in voluntary partnerships consistent with the principles of the United Nations Framework Convention on Climate Change (UNFCCC) (Singh and Mishra, 2015).

To further understand India's inherently equity-driven position, we must analyze the contemporary environmental challenges within their historical context, for which India advocates for developing mechanisms that acknowledge and rectify historical injustices in global environmental governance by challenging the dominant, neutral narratives advanced by developed countries. India argues for integrating a necessary historicity through principles like CBDR into international negotiations. Its approach has reshaped and transformed the global environmental discourse by focusing on the concerns and developmental needs of the Global South, which is the cornerstone of any dialogue on climate change concerns sensitive to the unique needs of developing countries (Stahlke, 2023). Through its advocacy, India emphasized the importance of balancing environmental sustainability with social and economic development, thereby contributing to the formulation of more inclusive and just international frameworks that are integral for the poorer nations and the poorest within them, who will be the worst hit by climate change (Stahlke, 2023). This is the fundamental concern at the heart of the climate change crisis. Those who have contributed the least to causing climate change will bear the real brunt of it, resulting in a violation of the human rights of its citizens within such jurisdictions.

The extent of application and enforcement of protected human rights will differ from state to state depending on national circumstances, constitutional culture, legislative approach, judicial activism, and governance mechanisms. Coincidentally, most parties to the core human rights treaties are also parties to the UNFCCC and other protocols and conventions. Within these states, many state parties are also parties to climate and human rights treaties. They are legally obliged to approach the climate change problem as a global environmental problem and a human rights issue (Upadhyay and Mehrotra, 2024). In this regard, the approach of equity and inclusion that India has argued and advocated for assumes significance as the developed states are obliged to identify and explore the human rights that might be placed at risk by climate impacts and assist through technology transfer, capacity building, and various other policies and measures to help the developing world mitigate and adapt to climate change, especially for protecting its citizens' economic, social, and cultural rights that are progressive and depend on the economic and social capacities of the states (Upadhyay and Mehrotra, 2024).

2.2 Climate Change and Human Rights Discourse in the Developing Countries

We have discussed the need to incorporate the views and concerns of developing countries, especially those from the Global South, within environmental governance. It becomes essential to discuss how the population within developing countries is affected by climate change, which violates their human rights. The impacts of climate change have severely undermined the livelihoods of vulnerable populations and exacerbated socioeconomic and political inequalities (Murcott and Tigre, 2024). Developing countries face the enormous challenges of lifting their citizens out of poverty and ensuring their basic needs, such as access to water, health, sanitation, education, and energy, while addressing the governance and developmental challenges. Due to these political reasons, the climate change issue is not given much attention. Developing countries also believe they have contributed relatively less to climate problems than industrialized developed countries. Hence, they should not be at the forefront of solving the climate problem. It is to be noted that climate change's impact will increasingly affect developing countries' development outcomes as they are mainly vulnerable to climate change. Climate change and changing weather patterns are already disrupting agricultural systems, water availability, forest health, and coastal ecology, affecting millions across the world. Hence, all countries should find ways to productively engage with the climate debate because there is no developmental path without addressing climate concerns (Chanduru, 2024). Thus, there is an urgent need to critically evaluate the intersection of climate change, human rights, and international law, focusing on implementing obligations toward developing states and the role of international treaties.

Climate change constitutes one of the most pressing global challenges, threatening the enjoyment of human rights by present and future generations. The United Nations General Assembly in 2022 recognized that unsustainable development and climate change represent serious threats to effectively realizing all human rights (Gowan, 2024). Moreover, those who are particularly vulnerable are marginalized groups and populations in the global South, who are disproportionately impacted due to their limited capacity to adapt and respond to climate-related challenges, which are compounded by systemic inequalities (Upadhyay and Mehrotra, 2024). These populations are often excluded from decision-making processes, leaving their economic, social, and cultural rights inadequately protected. In this regard, the states in the Global North, as the predominant contributors to climate change, bear distinctive obligations to address these inequities under principles of international law. To illustrate, the principles of responsibility under international law, such as "common but differentiated responsibilities and respective capabilities" (CBDR-RC), enshrined in Article 3 of the United Nations Framework Convention on Climate Change, emphasize the differentiated obligations of states in addressing climate change. Thus, the states in the Global North are required to take the lead in mitigation efforts and support developing countries for adaptation, capacity building, and financing. The causes and effects of climate change have elicited a broad response under international law. While the Paris Agreement (2015) is the primary legally binding instrument addressing climate change, there is growing recognition of international human rights treaties as a direct source of legal obligations in climate change.

These treaties mandate a holistic response encompassing adaptation, mitigation, loss and damage, financing, and education to safeguard human rights. Even the ongoing advisory opinion sought by the International Court of Justice (ICJ) further underscores the relevance of international human rights law in addressing climate change (UNGA, 2023). Some of the issues posed before the ICJ include states' legal obligation under international law regarding fossil fuels production under their jurisdiction, whether the object and purpose of the Paris Agreement and UNFCCC have any legal effects regarding the determination of whether Article 4 of the Paris Agreement (nationally determined contributions) provides only an obligation of conduct or also one of result; the legal content of the right to a clean, healthy and sustainable environment and its relationship with other human rights that states have considered relevant for this advisory opinion; and the legal significance of the declarations made by some states on becoming parties to the UN climate agreements to the effect that these agreements may not be interpreted as derogating from general principles of international law or any claims or rights concerning claiming compensation or liability (UNGA, 2023).

The need of the hour is for the ICJ to stipulate the human rights obligations under the UNFCCC to use human rights law and its related mechanisms to influence and improve international climate change policy governed by the UNFCCC to make that policy fairer and more effective. The questions that the ICJ must clarify include, but are not limited to, how human rights obligations and principles can be best applied to national climate change policy across both mitigation and adaptation, the precise nature of exterritorial human rights obligations as they relate to climate change, and how the international community, through the international human rights mechanisms, can best draw attention to and enforce those obligations in a manner that complements and supports the UNFCCC process.

Historically, in November 2007, the Small Island Developing States met in the Maldives and adopted the Male Declaration on the Human Dimension of Global Climate Change. The declaration laid a roadmap for actions within the UN system designed to explore and draw attention to the relationship between global warming and the full enjoyment of civil and political human rights, as well as economic, social, and cultural rights (Cameron, 2007). The questions that this declaration seeks to answer include how to understand the complex and multifaceted relationship between climate change and human rights, whether climate change constitutes a violation of human rights, especially the rights of vulnerable people, and what are states' national-level and international-level human rights obligations pertaining to climate change.

It needs to be understood that the reasons behind bringing the Male Declaration continue to exist even now, such reasons being frustration on the part of vulnerable communities at the slow pace of progress in tackling climate change, lack of emphasis on the victims around the world, and the lack of an accountability framework to deal with a phenomenon caused by man and with devastating human consequences.

This process highlights the need for developed states to assist developing countries in preventing violations, redressing existing harms, and establishing mechanisms to protect human rights from future climate-related risks. Furthermore, the role of Sustainable Development Goal (SDG) 13 calls for urgent action to combat climate change and its impacts (Arora and Mishra, 2023). This goal aligns with the broader objectives of the Paris Agreement and other international frameworks, emphasizing the interconnectedness of environmental sustainability and human rights. It needs to be understood that addressing climate change requires a robust, multidimensional approach that incorporates principles of equity, justice, and human rights as advocated by India and other developing countries. The obligations of states under international law, particularly those in the Global North, must reflect their historical responsibilities and capabilities. Thus, a comprehensive response, informed by the principles of CBDR-RC and guided by human rights treaties, is essential to safeguard the rights of vulnerable populations in developing countries to ensure a sustainable future.

From a comparative perspective, several countries have given due recognition to environmental rights in their constitutions, establishing respective liabilities of governments and corporations for environmental degradation, contributing to the consolidation and development of environmental jurisprudence, and providing a legal basis for individuals to benefit from these rights. To illustrate, South Africa's Constitution in 1996 gave legal recognition to environmental rights through section 24, which guarantees "everyone the right to an environment that is not harmful to their health or well-being" and calls on the government to "secure ecologically sustainable development." (Murcott and Vinti, 2024). Even Ecuador's Constitution in 2008 enshrined the concept of "Pachamama" (Mother Earth) and established mechanisms protecting ecosystems (Akchurin, 2015).

2.3 Need for Climate Justice: An Indian Perspective

Amid the background reflecting on the needs of developing countries, the term "climate justice" has gained significance, especially in the demands made by India on behalf of the various countries in the Global South. The climate justice movements originated in the early environmental protection movements that resulted in massive environmental pollution and health and other threats (Porter et al., 2020). The threat to climate systems is inherently global in nature, with most "perpetrators" often either not present anywhere near many of the worst-impacted areas, or not visible, or even outside national legal or other jurisdictions. Climate justice addresses the intersecting challenges of social, ecological, and economic inequities that climate change exacerbates. The Indian perspective on climate security highlights the interplay between developmental priorities and geopolitical pressures. For India, economic growth and development are seen as policy priorities, with the climate being the co-benefit in this movement toward a low-carbon future (Madhanagopal and Jacob, 2022).

On the domestic front, India's policies are driven by developmental priorities, while at the international level, its strategies are influenced by geopolitical pressures. These dual imperatives are not mutually exclusive but reflect a complex interplay of economic, environmental, and security considerations. According to this perspective, climate change is an environmental challenge and a profound social justice issue. It disproportionately impacts vulnerable communities, including indigenous groups and coastal populations, threatening their cultural identities and livelihoods. In this context, climate justice aims to ensure equitable access to the tools and knowledge necessary for these communities to adapt to the effects of climate change, which is a critical aspect of social justice (Madhanagopal and Jacob, 2022).

Historically, climate justice can be traced to the United Nations Conference on the Human Environment (1972), which sought to balance environmental protection with human development (Balboa, 1973). The Rio Declaration on Environment and Development (1992) further articulated climate justice principles known as the "Rio Principle"; it emphasized the need to protect people's rights through environmental sustainability and equitable development (Kovar, 1993). A critical example of the Rio Principle discussed above is the notion of "common but differentiated responsibilities and capabilities" (CBDR), which acknowledges the differing capacities of developed and developing nations in addressing climate crises. This principle underscores fairness and equity in burden-sharing, making it a cornerstone of climate justice discourse. Thus, integrating climate justice into local, national, and international policies is critical for addressing the disproportionate vulnerabilities of marginalized communities. These groups often lack the rights, resources, and preparedness to confront the adverse impacts of climate change, which makes climate change not solely an international issue but also a domestic one, emphasizing the need for inclusive and participatory approaches to climate governance. Back to the Indian perspective on climate change, historically strong environmental action has been taken, as demonstrated by the Bishnoi communities of Rajasthan, the "Chipko" Movement, the Silent Valley, and the Narmada movements (Abdul, 2017). Many issues range from environmental concerns and resistance movements around water, dams and rivers, forests, coal power plant pollution, and so on, which are prevalent in India. A key aspect of India's perspective is the recognition of the security implications of climate change. This is often framed regarding the threats posed to the country's economic development.

In the non-governmental realm, India's climate justice discourse is informed by a distinctive "environmentalism of the poor," which prioritizes the needs and rights of marginalized communities. Its principles, rooted in fairness and equity, demand the active participation of all stakeholders to ensure that the burdens and benefits of climate action are shared equitably. For countries like India, balancing developmental priorities with climate imperatives offers an opportunity to lead global efforts in advancing climate justice while securing domestic resilience. There has been a gradual process of integrating climate change into the domestic security policy-making agenda in the country, with the growing recognition of worsening climate vulnerabilities as Indian cities are especially vulnerable to climate change due to their rapid population growth, high levels of socioeconomic inequality, and the general inability of infrastructure and public services to adapt to projected impacts. These vulnerabilities include temperature variability, droughts, flooding, cyclones, sea-level rise, and the linked environmental health risks.

Given this context, specific steps have been taken by the Indian government at the national level, including the proposed amendments to India's Disaster Management Act. The proposed Disaster Management Amendment Bill (2024) aims to integrate disaster law with climate change adaptation and mitigation strategies (Nair et al., 2024). Key features include empowering the National and State Disaster Management Authorities (NDMA and SDMA) to prepare comprehensive disaster management plans. At the same time, further steps have been proposed, including establishing urban disaster management authorities for cities with municipal corporations, creating state disaster response forces (SDRF) for localized disaster preparedness and response, and mandating periodic audits of disaster preparedness and response measures. Lastly, steps have been proposed to strengthen the capacity of disaster management authorities, involving the development of national and state disaster databases to track risks, resource allocation, and preparedness, as well as providing technical assistance to local authorities and post-disaster audits to assess response effectiveness.

2.4 International Law, Climate Debate, India and Regional Arrangements

As discussed above, the evolving significance of climate change has long been realized in international and regional contexts. Starting in 1972, the Stockholm Declaration was widely regarded as the foundational document for environmental rights in international law. It asserts that humans have a "fundamental right" to a healthy environment, which is essential for their well-being and dignity. The Declaration laid the groundwork for subsequent international laws on environmental protection, many of which build on its twenty-six principles of environmental protection, sustainable development, and international cooperation in addressing environmental challenges. Subsequently, in 1992, the Rio Declaration on Environment and Development expanded on the principles enshrined in the Stockholm Declaration. For example, Principle 10 emphasizes the importance of public participation, access to information, and access to justice in environmental matters (Kovar, 1993). These principles are mainly attributed to the right to a clean environment, ensuring that individuals and communities have access to justice and law to protect their environmental rights.

The Rio Declaration emphasized the crucial integration of environmental protection into economic and social development by explicitly stating that "the right to development must be fulfilled so as to equitably meet developmental and environmental needs of present and future generations." Another significant milestone was achieved in 1992 when the UN Framework Convention on Climate Change (UNFCCC) was implemented. This convention has laid the foundation for collaborative efforts to address the dynamic and multifaceted issue of climate change (Hermwille et al., 2015). It recognized the need and urgency for global cooperation to mitigate and reduce the harm caused by human activities. Though the convention does not explicitly frame climate change as a human rights issue, it establishes the principles of equity and sustainable development, which are closely linked to the right to a healthy environment. As a follow-up, the Kyoto Protocol was adopted on December 11, 1997, which operationalizes the United Nations Framework Convention on Climate Change by committing industrialized countries and economies in transition to limit and reduce greenhouse gas (GHG) emissions in accordance with agreed individual targets (Böhringer, 2003).

After a gap of more than a decade since the Kyoto Protocol, it was in 2015 that the Paris Agreement on Climate Change, which is the most important international instrument on environmental protection, was established (Savaresi, 2016). Its Preamble stresses that climate change presents a significant challenge to the enjoyment of human rights, including the right to health, the right to food, water, and an adequate standard of living. It also lays down standards that commit nations to limit the global temperature increase to below 2 degrees Fahrenheit above the pre-industrial threshold and to make efforts to limit the temperature increase to a maximum of 1.5 degrees Fahrenheit.

As discussed above, India has played a pivotal role in developing the international climate change regime, particularly by formulating the foundational principle of common but differentiated responsibilities (CBDR). This principle has ensured a clear differentiation between developed and developing countries within the framework of the United Nations Framework Convention on Climate Change (UNFCCC) and its Kyoto Protocol. India's contribution was instrumental in drafting these key legally binding instruments, including the UNFCCC and its Kyoto Protocol. Notably, India presented the first full draft of the UNFCCC for negotiation and adoption by member states.

At the same time, India also introduced an alternative, deliberate, and thoughtful vision for international climate change law. This vision proposed intermingling soft and hard laws, blending formal legal instruments with informal normative frameworks. In this approach, India emphasized the importance of integrating domestic and international legal benchmarks to establish meaningful "bindingness" within the evolving climate change regime. This approach reflected a nuanced understanding of global climate governance, balancing equity, development needs, and environmental sustainability.

Similar to the approach propounded by India, several Regional Human Rights arrangements have played a crucial role in channelling and codifying the right to a clean, safe, and healthy environment that has put forth the concerns of the developing world. Some such vital instruments include the African Charter on Human and Peoples' Rights· Article 24 explicitly recognizes the right of all peoples to a generally satisfactory environment favorable to their development. This provision has been utilized manifold as the fundamental basis to initiate legal challenges against the ever-increasing environmental degradation in multiple African countries, contributing immensely to the growth and development of environmental jurisprudence in Africa. Similarly, the Additional Protocol to the American Convention on Human Rights in the Area of Economic, Social, and Cultural Rights (Protocol of San Salvador) in 1988 enshrined that "Everyone shall have the right to live in a healthy environment and to have access to basic public services." (LeBlanc, 1992). This regional instrument has greatly influenced the development and consolidation of environmental law in Latin America.

2.5 India and Climate Change: Lessons to Learn for India

The issues surrounding climate change hold specific relevance for developing countries, including India, as discussed in this article, since these countries have encountered significant difficulties in tackling climate change both historically and economically. Climate change further substantiates this by presenting unparalleled threats to human life, property, and ecosystems. Despite growing recognition of these challenges, existing international legal and policy frameworks remain fragmented, particularly concerning disaster response and climate justice. However, nations can enhance disaster resilience and ensure climate justice by integrating human rights-based approaches and promoting inclusive decision-making (Upadhyay and Mehrotra, 2024).

In this context, India's emphasis as a representative of developing countries has been that development should be inclusive, equitable, and people-centered, advocating for international cooperation to support the economic growth of underdeveloped regions. This perspective is crucial for developing countries as it perceives development as a matter of right, not a privilege. The right to development pushes for equitable redistribution of global resources, fair trade practices, and international support for poverty alleviation, all of which are essential to fostering sustainable progress in the Global South and for overall well-being and sustainability. Climate change threatens small island states and coastal communities in the Global South, intensifying the need for robust environmental protections. This concern is further aggravated by the absence of a comprehensive international legal framework for disaster response, creating significant gaps in ensuring climate justice and human rights protection. While regional laws and bilateral agreements provide partial guidance, they fail to address global challenges comprehensively. The urgent need for a unified legal framework is underscored by the necessity to enhance disaster response and risk reduction capabilities at both international and national levels.

Thus, the increasing reliance on principles like common but differentiated responsibilities (CBDR) that India has advocated can be seen as a part of the larger picture of reimagining international law to identify, address, and eradicate the historical injustices and negative power dynamics developing countries face. Traditional international law has repeatedly reiterated the aspirations and priorities of the Global North, emphasizing individual rights and economic and legal frameworks that benefit almost all developed nations. However, it is pertinent to mention that principles like CBDR challenge this structure by advocating for a more inclusive and balanced legal framework that understands and recognizes the unique and specific needs of the Global South. This shift aligns international law with the principles of equity, fairness, and shared responsibility, acknowledging that global prosperity requires lifting all regions out of poverty and fostering sustainable development. Consequently, it is expected to encourage the Global North to accept accountability for its environmental and economic impact on the Global South and support initiatives that mitigate these effects, including technology transfer, climate financing, and capacity building.

As we continue to embark on the journey to combat climate change, the challenge lies in bridging the divide between the Global North and Global South to achieve a climate regime that truly reflects the needs and aspirations of all humanity, especially for the Global South. Since the global climate change challenges continue to intensify, the steps taken by countries like India at the municipal level will be further scrutinized by the developed countries, who are pressured to take up their responsibilities at the global level. Thus, in addition to demanding cooperation from the developed countries, India, as a responsible member state, must also put in place certain measures that reflect its commitment to tackling climate change.

Some of the issues India needs to address include the lack of systemic integration of legal expertise into climate policy and disaster management frameworks, which has hindered efforts to address climate change effectively. By integrating human rights-based approaches and ensuring inclusive decision-making, India must advocate for a comprehensive legal framework at the international level to strengthen disaster resilience and promote equitable climate action for the welfare of developing countries.

However, India must consider its shortcomings and challenges in addressing climate change and disaster risks. Key issues include limited resources, low bureaucratic stability, and a disconnect between technical climate knowledge and environmental risk experiences. The lack of systemic integration of legal expertise further exacerbates these challenges. This can be seen in India's international negotiations on climate change that often lack consistent legal consultation; the Legal and Treaties Division of the Ministry of External Affairs is consulted sporadically rather than systematically (Mehrotra and Upadhyay, 2024). This highlights the need for institutional mechanisms to ensure the integration of legal expertise in drafting submissions and developing international legal texts. Furthermore, the absence of robust adaptation policies has resulted in violations of social and economic rights for populations residing in affected regions. Some reasons behind the failure of adaptation and resilience actions in India are the limited autonomy of local governance structures, low awareness, technical capacity, and siloed approaches to climate action.

The need of the hour is to implement effective adaptation strategies that foresee the consequences of climate change for humans, both as individuals and as part of communities. To put such policies in place, India needs to take certain steps, including promoting radical and progressive visions of climate justice through a multidimensional approach, which must include four critical dimensions. The first such dimension includes addressing differential climate impacts, that is, the effect of climate change on urban poor communities that are disproportionately exposed to climate risks due to inadequate poverty alleviation programs, social exclusion, and gaps in capacity building. Recognizing these differential impacts may result in equitable climate policies. Second, the need to tackle the root causes of vulnerability, which means efforts made to achieve climate justice, shall address the legacy of unequal development, exclusion, and dispossession in several parts of India. Thus, tackling these systemic issues is fundamental to realizing equitable outcomes. Thirdly, inclusive decision-making must delineate shared responsibilities among stakeholders, including the regional and central governments, and foster partnerships between state and non-state actors to promote accountability and collective action. Lastly, an intersectional approach to climate justice must be adopted that must address structural inequalities based on gender, class, caste, ethnicity, and citizenship status. Such an approach ensures that the voices of marginalized groups are integrated into climate action plans.

2.6 Conclusion

India's engagement in global climate diplomacy reflects its dual commitment to climate action and economic development. While its stance on equity and differentiated responsibilities has sometimes drawn criticism, India remains critical in shaping international climate governance. As climate vulnerabilities worsen, India's ability to balance domestic priorities with its global commitments will be crucial for the success of both national and international climate objectives. India is poised to play a pivotal role in the global fight against climate change by advocating for climate justice and leading sustainable development initiatives.

At the domestic level, India's environmental governance framework reflects a strong commitment to justice, equity, and developmental priorities. By aligning its National Environmental Policy with the principles of CBDR and the Right to Development, India seeks to address structural inequities in global environmental governance. Its reliance on multilateralism and voluntary partnerships underscores a cooperative approach that prioritizes the interests of the Global South while contributing to sustainable global solutions. This strategic philosophy positions India as a leading advocate for equitable and inclusive climate action, which seeks to address the interconnected challenges of social, ecological, and economic inequities that are exacerbated by climate change. From an Indian perspective, the concept of climate security underscores the delicate balance between developmental priorities and geopolitical pressures, reflecting the complex interplay of national interests and global environmental responsibilities.

To facilitate India's meaningful participation in the international climate regime, it is essential to understand the country's moral and material leadership aspirations amid pressing domestic concerns. India, having contributed relatively little to the climate problem, is deeply vulnerable to its impacts in achieving its developmental goals. Hence, India's climate diplomacy must navigate the complex intersection of environmental responsibility, economic development, and increased productivity while actively finding ways to mitigate and adapt to climate change. India must reconcile its historical grievances and use soft power to address the urgent need for climate action under SDG 13. This will involve crafting policies that address the inequities of the past while ensuring sustainable development and human rights for all.

REFERENCES

Akchurin, M. (2015). Constructing the rights of nature: Constitutional reform, mobilization, and environmental protection in Ecuador. *Law & Social Inquiry*, *40*(4), 937–968. https://doi.org/10.1111/lsi.12141

Arora, N. K., & Mishra, I. (2023). Sustainable development goal 13: Recent progress and challenges to climate action. *Environmental Sustainability*, 6, 297–301. https://doi.org/10.1007/s42398-023-00287-4

Balboa, Marie Whitesell. (1973). United nations conference on the human environment. *Women Lawyers Journal*, 59(1), 26-33.

Böhringer, C. (2003). The Kyoto Protocol: A review and perspectives. *Oxford Review of Economic Policy*, *19*(3), 451–466. https://doi.org/10.1093/oxrep/19.3.451

Cameron, E. (2007). *Declaration on the human dimension of global climate change*. Center for International Environmental Law (CIEL).

Chanduru, K. R. (2024). Climate change and the right to development: Legal perspectives and challenges. *Nyaayshastra Law Review*, *4*(2), 1–15.

Dubash, N. K. (Ed.). (2019). An introduction to India's evolving climate change debate: From diplomatic insulation to policy integration. In *India in a warming world: Integrating climate change and development* (pp. 1–28). Oxford Academic.

Gowan, R. (2024). The United Nations General Assembly's place in crisis response. *Fletcher Forum of World Affairs*, *48*(2), 1–8.

Gupta, H., Kohli, R. K., & Ahluwalia, A. S. (2015). Mapping 'consistency' in India's climate change position: Dynamics and dilemmas of science diplomacy. *Ambio*, *44*, 592–599. https://doi.org/10.1007/s13280-014-0609-5

Hermwille, L., Obergassel, W., Ott, H. E., & Beuermann, C. (2015). UNFCCC before and after Paris – what's necessary for an effective climate regime? *Climate Policy*, *17*(2), 150–170. https://doi.org/10.1080/14693062.2015.1115231

Kovar, J. D. (1993). A short guide to the Rio Declaration. *Colorado Journal of International Environmental Law and Policy*, *4*(1), 119–140.

LeBlanc, L. (1992). The economic, social, and cultural rights protocol to the American Convention and its background. *Netherlands Quarterly of Human Rights*, *10*(2), 130–154. https://doi.org/10.1177/016934419201000203

Madhanagopal, D., & Jacob, V. A. (2022). Climate risks in an unequal society: The question of climate justice in India. In D. Madhanagopal, C. T. Beer, B. R. Nikku, & A. J. Pelser (Eds.), *Environment, climate, and social justice* (pp. 123–140). Springer. https://doi.org/10.1007/978-981-19-1987-9_9

Mann H (1992). The Rio Declaration. Proceedings of the ASIL Annual Meeting; 86: 405–414. doi:10.1017/S0272503700095367

Mehrotra, A., & Upadhyay, A. (2024). India's approach towards adjudication before the International Court of Justice and International Criminal Court: In search of uniformity. *Janus.net, e-Journal of International Relations*, *15*(2), 295–308. https://doi.org/10.26619/1647-7251.15.2.13

Murcott, M. J., & Tigre, M. A. (2024). Developments, opportunities, and complexities in global south climate litigation: Introduction to the special collection. *Journal of Human Rights Practice*, *16*(1), 1–24. https://doi.org/10.1093/jhuman/huad070

Murcott, M. J., & Vinti, C. (2024). The judge-made 'duty' to consider climate change in South Africa. *Journal of Human Rights Practice*, *16*(1), 125–145. https://doi.org/10.1093/jhuman/huad069

Nair, S. S., Singh, S., & Gupta, A. K. (2024). Climate adaptation and disaster risk reduction integration through environmental legislation in India. In A. K. Gupta, A. Gupta, & P. Acharya (Eds.), *Disaster risk and management under climate change* (pp. 165–180). Springer. https://doi.org/10.1007/978-981-99-4105-6_22

Nazer.K, A. (2017). THE ENVIRONMENTAL MOVEMENTS IN INDIA VIS. A VIS. CONCEPT OF DEVELOPMENT: A PARADIGM SHIFT FROM NEHRUVIAN MODEL TO GANDHIAN MODEL. *The Indian Journal of Political Science*, 78(4), 493–497.

Porter, L., Rickards, L., Verlie, B., Bosomworth, K., Moloney, S., Lay, B., & Pellow, D. (2020). Climate justice in a climate-changed world. *Planning Theory & Practice*, *21*(2), 293–321. https://doi.org/10.1080/14649357.2020.1748959

Savaresi, A. (2016). The Paris Agreement: A new beginning? *Journal of Energy & Natural Resources Law*, *34*(1), 16–26. https://doi.org/10.1080/02646811.2016.1133983

Singh, R. K. & Mishra, A. (2015). Revisiting India's environmental policy. *The Indian Journal of Political Science*, *76*(3), 639–646.

Stahlke, T. (2023). Climate policy and the concept of co-benefits in India. *Journal of Socio-Economic Development*, *25*, 86–102. https://doi.org/10.1007/s40847-023-00235-2

UNGA. (2023). *Request for an advisory opinion of the International Court of Justice on the obligations of states in respect of climate change*. United Nations.

Upadhyay, A., & Mehrotra, A. (2024). Need for a human rights-based approach to climate change adaptation and mitigation with reference to the state of Uttarakhand in India. In Ş. N. Açikalin & Ş. Ş. Erçetin (Eds.), *The social consequences of climate change* (pp. 67–79). Emerald Publishing Limited. https://doi.org/10.1108/978-1-83797-677-520241006

World Bank Group. (2020, June 8). *COVID-19 to plunge global economy into worst recession since World War II*. Press release No: 2020/209/EFI. https://www.worldbank.org/en/news/press-release/2020/06/08/covid-19-to-plunge-global-economy-into-worst-recession-since-world-war-ii

3

Climate Diplomacy and Small Island Developing States (SIDS)

Mohd Imran & Aishath Shafeeu

3.1. Introduction

Climate diplomacy is thestrategic engagement between states, international organizations, and other actors to address the global challenges posed by climate change. It seeks to build consensus, create alliances, and develop comprehensive strategies to mitigate the impacts of climate change while promoting sustainable development. Climate diplomacy has become a critical tool for Small Developing Island States (SIDS) in raising concerns about the effects of climate change that threaten the resources, economies, and livelihoods of these nations.

SIDS stand on the frontlines of the global climate crisis, bearing the worst of its impacts despite contributing the least to its causes. These nations, scattered across the Pacific, Caribbean, and Indian Ocean, are characterized by their small habitable land areas, limited resources, and low economic diversification. However, their geographic and economic vulnerabilities make them disproportionately susceptible to the devastating effects of climate change, including erratic weather events, rising sea levels, and loss of biodiversity.

The vulnerability of SIDS to climate change is exacerbated by their limited capacity to adapt and recover from its impacts. With economies heavily reliant on climate-sensitive sectors such as tourism, fishing, and agriculture, SIDS face significant challenges in building climate resilience and achieving sustainable development. Moreover, their comparatively small populations, weak economies, and unique geographical isolation mean their voices are often marginalized in global climate negotiations, despite their authority as the world's most climate-vulnerable nations. Yet, SIDS have emerged as powerful advocates for global action, utilizing their experiences to lobby for ambitious targets, equitable climate finance mechanisms, and international cooperation in achieving climate goals. Through coalitions such as the Alliance of Small Island States (AOSIS), creative advocacy strategies, and regional partnerships, SIDS have played a pivotal role in creating and shaping international climate policies, including the landmark Paris Agreement in 2016. However, SIDS continue to face significant challenges, including limited resources, weak administration, and capacity constraints to geopolitical disparities that hinder their ability to secure the climate funding required to protect their land and people from the ever-growing effects of climate change.

This paper explores the multifaceted vulnerabilities faced by SIDS to climate change and examines their critical role in global climate diplomacy while highlighting the pathways through which they can strengthen their advocacy in the international climate arena. By analyzing the complex interplay of environmental, economic, and geopolitical factors, it highlights the pressing need for collective action by the international community to address the challenges faced by SIDS and ensure their survival in a rapidly changing world.

3.2. The Vulnerability of SIDS to Climate Change

Small Island Developing States (SIDS) are among the most vulnerable regions in the world to the impacts of climate change. Their unique geographic, economic, and ecological characteristics make them particularly susceptible to the adverse effects of rising global temperatures, sea-level rise, extreme weather events, and biodiversity loss. These challenges are compounded by their limited resources, small economies, and dependence on fragile ecosystems, which exacerbate their vulnerability and hinder their ability to adapt effectively. The socioeconomic impacts of climate change on SIDS communities are profound, threatening livelihoods, cultural heritage, and the very existence of some nations.

SIDS are predominantly located in the tropics and subtropics, spanning the Caribbean, the Pacific, and the Indian Ocean. Their geographic isolation, small land areas, and limited natural resources make them inherently vulnerable to external shocks, including those induced by climate change. Many SIDS are low-lying islands or atolls, with some nations, such as the Maldives and Kiribati, having an average elevation of just a few meters above sea level. This makes them particularly susceptible to sea-level rise, which is projected to accelerate in the coming decades due to global warming (IPCC, 2023). For example, the

Intergovernmental Panel on Climate Change (IPCC) estimates that global sea levels could rise by up to 1.1 meters by 2100 under high-emission scenarios, which would inundate large portions of these islands and render some uninhabitable (IPCC, 2023).

Economically, SIDS face significant challenges due to their small size, limited diversification, and dependence on a narrow range of industries, such as tourism, fisheries, and agriculture. These sectors are highly sensitive to climate change, making SIDS economies disproportionately vulnerable. For instance, coral reef degradation caused by ocean warming and acidification threatens fisheries and tourism, which are critical sources of income and employment for many SIDS (Barros et al., 2014). Additionally, the high costs of importing goods and services due to their remote locations further strain their economies, leaving them with limited fiscal space to invest in climate adaptation and resilience-building measures (Briguglio, 2016).

Sea-level rise is one of the most immediate and existential threats faced by SIDS. As global temperatures rise, thermal expansion of seawater and the melting of polar ice caps contribute to rising ocean levels, which encroach on coastal areas, erode shorelines, and contaminate freshwater resources through saltwater intrusion. For example, in the Pacific island nation of Tuvalu, saltwater intrusion has already compromised agricultural lands and freshwater supplies, threatening food security and public health (Kelman, 2014). Similarly, in the Caribbean, rising sea levels exacerbate coastal flooding during storms, damaging infrastructure and displacing communities (Mycoo, 2018). Extreme weather events, such as hurricanes, cyclones, and typhoons, are becoming more frequent and intense due to climate change, posing significant risks to SIDS. These events can cause widespread destruction, including loss of life, damage to homes and infrastructure, and disruption of essential services. For instance, Hurricane Maria in 2017 devastated Dominica, causing losses equivalent to 226 percent of the country's GDP (World Bank, 2018). In the Maldives, the 2004 Indian Ocean Tsunami seriously affected the livelihoods of nearly a third of the population and caused losses equivalent to 62 percent of the national GDP (Asian Development Bank, 2005). Such events not only result in immediate humanitarian crises but also have long-term economic consequences, as resources that could have been used for development are diverted to disaster recovery and reconstruction.

Biodiversity loss is another critical issue for SIDS, as many of these nations are home to unique and fragile ecosystems, including coral reefs, mangroves, and rainforests. These ecosystems provide essential services, such as coastal protection, food security, and tourism revenue, but they are under increasing threat from climate change. Coral reefs, for example, are highly sensitive to rising sea temperatures and ocean acidification, which cause coral bleaching and reduce their ability to support marine life. The loss of coral reefs not only affects fisheries but also diminishes the natural barrier they provide against storm surges and coastal erosion (Hoegh-Guldberg, 2017). Similarly, mangrove forests, which act as natural buffers against rising seas and extreme weather, are being degraded due to human activities and climate change, further exacerbating the vulnerability of SIDS communities (Alongi, 2015).

The socioeconomic impacts of climate change on SIDS communities are multifaceted and far-reaching. One of the most pressing concerns is the threat to human security and displacement. As sea levels rise and extreme weather events become more frequent, many SIDS face the prospect of forced migration and the loss of habitable land. For example, in Kiribati, the government has explored options for relocating its population to other countries as a last resort to address the existential threat posed by climate change (Farbotko & Lazrus, 2012). In 2023, Australia signed the Falepili Union Treaty, which offered refuge to Tuvalu citizens grappling with the prospect of the country becoming uninhabitable as a result of rising sea levels (Green & Guilfoyle, 2024). This raises complex legal and ethical questions about the rights of climate refugees and the responsibilities of the international community to provide assistance and protection. Climate change also exacerbates existing inequalities and vulnerabilities within SIDS communities. Women, children, the elderly, and marginalized groups are often disproportionately affected due to their limited access to resources, information, and decision-making processes. For instance, women in SIDS are frequently responsible for water collection and food production, tasks that become increasingly difficult as freshwater sources are compromised and agricultural yields decline due to climate change (Nelson et al., 2002). Similarly, children are more susceptible to health risks, such as malnutrition and vector-borne diseases, which are expected to increase as temperatures rise and weather patterns become more erratic (McMichael et al., 2006).

The cultural heritage of SIDS is also under threat from climate change. Many SIDS have rich traditions and ways of life that are closely tied to their natural environments. In the Pacific, traditional practices such as fishing, farming, and navigation are deeply embedded in the cultural identity of island communities. However, as climate change alters ecosystems and disrupts these practices, there is a risk of cultural erosion and loss of traditional knowledge (Mercer et al., 2010). This not only affects the social fabric of SIDS communities but also undermines their resilience, as traditional knowledge often plays a crucial role in adapting to environmental changes.

The vulnerability of SIDS to climate change is a multifaceted issue that encompasses geographic, economic, ecological, and socioeconomic dimensions. Rising sea levels, extreme weather events, and biodiversity loss pose existential threats to these nations, while the socioeconomic impacts exacerbate existing vulnerabilities and inequalities. Addressing these challenges requires urgent and concerted action at the local, regional, and global levels, including enhanced climate adaptation measures, increased financial and technical support, and stronger international cooperation to mitigate the root causes of climate change. Without such efforts, the future of SIDS and their communities remains uncertain.

3.3. The Role of Climate Diplomacy in SIDS Advocacy

Small Island Developing States (SIDS) have emerged as some of the most vocal and influential advocates for global climate action. Despite their limited size and resources, these nations have played a pivotal role in shaping international climate negotiations, leveraging their moral authority and unique vulnerabilities to push for stronger commitments to mitigate climate change and support adaptation efforts. Through the formation of coalitions, strategic diplomacy, and innovative advocacy strategies, SIDS have amplified their concerns on international platforms, ensuring that the voices of those most affected by climate change are heard and addressed.

SIDS have long been at the forefront of global climate negotiations, using their firsthand experiences with the devastating impacts of climate change to advocate for urgent and ambitious action. Their moral authority stems from the fact that they are among the countries least responsible for global greenhouse gas emissions yet bear the brunt of climate change impacts. SIDS collectively account for less than 1 percent of global carbon emissions, yet they face existential threats from rising sea levels, extreme weather events, and ecosystem degradation (Thomas et al., 2020). This disparity has positioned SIDS as powerful advocates for climate justice, emphasizing the need for developed nations to take responsibility for their historical emissions and provide financial and technical support to vulnerable countries. One of the most significant contributions of

SIDS to global climate negotiations has been their role in advocating for the inclusion of the 1.5 degrees Celsius temperature limit in the Paris Agreement. Recognizing that even a 2 degrees Celsius rise in global temperatures would have catastrophic consequences for their survival, SIDS successfully lobbied for the more ambitious target, which was ultimately adopted as part of the agreement (Robinson, 2020). This achievement underscores the ability of SIDS to influence global climate policy despite their limited economic and political power. Their advocacy has also highlighted the importance of addressing loss and damage, a critical issue for SIDS that refers to the irreversible impacts of climate change that cannot be mitigated or adapted to (Vanhala & Hestbaek, 2016).

The formation of coalitions has been a key strategy for SIDS to amplify their voices and enhance their influence in international climate negotiations. The Alliance of Small Island States (AOSIS), established in 1990, is one of the most prominent examples of such a coalition. AOSIS represents thirty-nine member states and five observer states, including SIDS from the Caribbean, Pacific, and Indian Ocean regions, and has been instrumental in advocating for the interests of small island nations in global climate forums (Betzold, 2010). By uniting under a common platform, SIDS have been able to pool their resources, share expertise, and present a cohesive and compelling case for climate action. AOSIS has played a critical role in shaping key international climate agreements, including the United Nations Framework Convention on Climate Change (UNFCCC), the Kyoto Protocol, and the Paris Agreement. For example, during the negotiations leading up to the Paris Agreement, AOSIS was a leading voice in advocating for the inclusion of the 1.5 degrees Celsius target, as well as provisions for loss and damage and climate finance (Deitelhoff & Wallbott, 2012). The coalition's ability to build alliances with other negotiating groups, such as the least developed countries (LDCs) and the European Union, has further strengthened its influence in global climate diplomacy (Khan et al., 2020).

In addition to AOSIS, SIDS have formed regional coalitions to address specific climate-related challenges. For instance, the Pacific Islands Forum (PIF) has been active in advocating for stronger climate action and resilience-building measures in the Pacific region. Similarly, the Caribbean Community (CARICOM) has worked to coordinate climate policies and strategies among its member states, emphasizing the need for regional cooperation to address shared vulnerabilities (Pelling & Uitto, 2001). These coalitions have enabled SIDS to leverage their collective strength and ensure that their concerns are prioritized in international climate discussions.

SIDS have employed a range of innovative strategies to amplify their concerns and advocate for climate action on international platforms. One of the most effective strategies has been the use of storytelling and personal narratives to highlight the human impacts of climate change. By sharing firsthand accounts of how rising sea levels, extreme weather events, and ecosystem degradation are affecting their communities, SIDS have been able to humanize the climate crisis and galvanize support for their cause (Farbotko & Lazrus, 2012). For example, the former president of the Maldives, Mohamed Nasheed, famously held an underwater cabinet meeting in 2009 to draw attention to the threat of sea-level rise to his nation's survival (Randall, 2011). Such creative and compelling advocacy efforts have helped to raise awareness of the unique challenges faced by SIDS and build momentum for global climate action (Ananthavinayagan et al., 2024).

Another key strategy has been the use of scientific evidence and data to support their advocacy efforts. SIDS have collaborated with researchers and international organizations to produce reports and studies that document the impacts of climate change on their nations and highlight the urgent need for action. For instance, the IPCC's Special Report on Global Warming of 1.5 degrees Celsius, published in 2018, was heavily influenced by the advocacy of SIDS and provided scientific evidence to support their call for more ambitious climate targets (IPCC, 2018). By grounding their advocacy in robust scientific evidence, SIDS have been able to strengthen their credibility and persuade other nations to take their concerns seriously. SIDS have also leveraged international legal frameworks to advance their climate advocacy. For example, in 2021, the Pacific island nation of Tuvalu announced its intention to explore legal options for holding major polluters accountable for climate change under international law (Farbotko, 2021). Similarly, SIDS have been active in advocating for the recognition of climate refugees and the development of legal frameworks to protect the rights of those displaced by climate change (McAdam, 2012). These efforts reflect the determination of SIDS to use all available tools to address the climate crisis and ensure that their voices are heard.

In addition to these strategies, SIDS have sought to build partnerships with non-state actors, including civil society organizations, the private sector, and youth groups, to amplify their advocacy efforts. For instance, the Pacific Islands Climate Action Network (PICAN) has worked closely with SIDS governments to mobilize grassroots support for climate action and advocate for stronger policies at the international level (Kelman, 2014). Similarly, SIDS have engaged with the private sector to promote sustainable development and climate resilience, recognizing the importance of collaboration in addressing the complex challenges posed by climate change (Nunn et al., 2016).

The role of climate diplomacy in SIDS advocacy cannot be overstated. Through their leadership in global climate negotiations, the formation of coalitions such as AOSIS, and the use of innovative strategies to amplify their concerns, SIDS have made significant contributions to advancing global climate action. Their efforts have not only raised awareness of the unique challenges faced by small island nations but also underscored the urgent need for collective action to address the climate crisis. As the impacts of climate change continue to intensify, the advocacy of SIDS will remain critical in ensuring that the voices of the most vulnerable are heard and that the international community takes meaningful steps to protect their future.

3.4. Challenges in Climate Diplomacy for SIDS

While Small Island Developing States (SIDS) have been influential advocates for global climate action, they face significant challenges in their diplomatic efforts. These challenges stem from their limited resources and capacity for negotiation, the need to balance development goals with climate resilience, and the geopolitical disparities that often marginalize their voices in international climate decision-making. Addressing these obstacles is critical for ensuring that SIDS can effectively advocate for their interests and secure the support they need to build climate resilience and achieve sustainable development.

One of the most significant challenges facing SIDS in climate diplomacy is their limited financial, technical, and human resources. Many SIDS have small populations and economies, which constrain their ability to maintain large diplomatic missions or specialized teams for climate negotiations. For example, some Pacific island nations have only a handful of representatives at major international forums, making it difficult for them to participate effectively in multiple simultaneous negotiations or technical discussions (Betzold, 2010). This lack of capacity often puts SIDS at a disadvantage compared to larger, wealthier nations that can deploy extensive resources and expertise to advance their interests. The financial constraints faced by SIDS also limit their ability to invest in climate adaptation and mitigation measures, which are critical for building resilience and reducing vulnerability. Many SIDS rely heavily on external funding, such as grants and loans from international organizations and donor countries, to finance their climate initiatives. However, accessing these funds can be a complex and time-consuming process, requiring significant administrative capacity and technical expertise (Nurse et al., 2014). Moreover, the funds available are often insufficient to meet the scale of the challenges faced by SIDS, leaving them with limited options for addressing the impacts of climate change.

In addition to financial and technical constraints, SIDS often face challenges in building and retaining the expertise needed for effective climate diplomacy. Many SIDS lack specialized training programs or institutions focused on climate change, making it difficult to develop a pool of skilled negotiators and policymakers (Thomas et al., 2020). This shortage of expertise can hinder their ability to engage in complex technical discussions, such as those related to climate finance mechanisms or the implementation of the Paris Agreement. Furthermore, the brain drain phenomenon, where skilled professionals emigrate to larger countries in search of better opportunities, exacerbates the capacity challenges faced by SIDS (Connell, 2013).

Another major challenge for SIDS is balancing their development goals with the need to build climate resilience. Many SIDS are classified as middle-income countries, which limits their access to concessional financing and grants for climate adaptation and mitigation. At the same time, they face significant development challenges, including poverty, unemployment, and inadequate infrastructure, which require substantial investments (Mycoo, 2018). This creates a dilemma for SIDS, as they must allocate limited resources between immediate development needs and long-term climate resilience. For example, tourism is a critical sector for many SIDS, contributing significantly to GDP and employment. However, tourism is highly vulnerable to climate change, as rising sea levels, extreme weather events, and coral reef degradation can damage infrastructure and deter visitors (Scott et al., 2012).

Investing in climate-resilient infrastructure and sustainable tourism practices is essential for protecting this sector, but such investments often compete with other development priorities, such as healthcare and education. Similarly, agriculture, which is a key source of livelihood for many SIDS communities, is increasingly threatened by climate change, requiring investments in climate-resilient crops and farming practices (Mimura et al., 2007). The challenge of balancing development and climate resilience is further complicated by the need to transition to low-carbon economies. Many SIDS rely heavily on imported fossil fuels for energy, which not only contributes to greenhouse gas emissions but also exposes them to volatile global energy prices (Dornan, 2014). Transitioning to renewable energy sources, such as solar, wind, and geothermal, can enhance energy security and reduce emissions, but it requires significant upfront investments and technical expertise. For SIDS with limited financial resources, this transition poses a significant challenge, particularly when competing with other pressing development needs.

Geopolitical disparities present another major challenge for SIDS in climate diplomacy. Despite their moral authority and advocacy efforts, SIDS often find themselves marginalized in international climate negotiations, where power dynamics are heavily skewed in favor of larger, wealthier nations. This marginalization is evident in the unequal distribution of decision-making power within international institutions, such as the United Nations Framework Convention on Climate Change (UNFCCC), where consensus-based decision-making can disadvantage smaller nations (Khan et al., 2020). For example, SIDS have frequently expressed frustration over the slow pace of negotiations and the reluctance of major emitters to commit to ambitious climate targets or provide adequate financial support for adaptation and loss and damage (Robinson, 2020).

The influence of geopolitical interests on climate decision-making is another significant barrier for SIDS. Major emitters and industrialized nations often prioritize their economic and political interests over the needs of vulnerable countries, leading to weak or inadequate climate agreements. For instance, the failure of the Copenhagen Climate Summit in 2009 to deliver a binding agreement was widely attributed to the reluctance of major emitters to accept stringent emission reduction targets (Depledge, 2010). This outcome was a significant setback for SIDS, which had been advocating for stronger commitments to limit global warming to 1.5 degrees Celsius. Moreover, the allocation of climate finance often reflects geopolitical biases, with a disproportionate share of funds directed toward middle-income countries or regions with strategic importance to donor nations (Weikmans & Roberts, 2019). This leaves SIDS with limited access to the resources they need to implement climate adaptation and mitigation measures. For example, the Green Climate Fund (GCF), which was established to support developing countries in addressing climate change, has been criticized for its complex and bureaucratic application process, which disproportionately disadvantages smaller nations with limited administrative capacity (Betzold & Weiler, 2017).

In addition to these challenges, SIDS must navigate the complexities of international climate governance, which often involves multiple overlapping agreements and institutions. Coordinating their positions across these forums requires significant diplomatic effort and resources, which can strain the limited capacity of SIDS. For example, SIDS must engage with not only the UNFCCC but also other international bodies, such as the International Maritime Organization (IMO) and the International Civil Aviation Organization (ICAO), which regulate emissions from shipping and aviation, respectively (Kelman, 2014). This fragmentation of climate governance can create inconsistencies and gaps in addressing the specific needs of SIDS.

The challenges faced by SIDS in climate diplomacy are multifaceted and deeply rooted in their unique vulnerabilities and the broader geopolitical context. Limited resources and capacity for negotiation, the need to balance development goals with climate resilience, and the marginalization of their voices in international decision-making all pose significant obstacles to their efforts to advocate for stronger climate action. Addressing these challenges requires a concerted effort by the international community to provide SIDS with the financial, technical, and political support they need to build resilience and achieve sustainable development. Without such support, the ability of SIDS to adapt to the impacts of climate change and secure their future will remain severely constrained.

3.5. Pathways for Strengthening Climate Diplomacy for SIDS

Small Island Developing States (SIDS) face unique challenges in their efforts to advocate for climate action and build resilience to the impacts of climate change. However, there are several pathways through which SIDS can strengthen their climate diplomacy and enhance their capacity to address these challenges. These include enhancing international partnerships and funding mechanisms, leveraging technology and data for evidence-based advocacy, and building long-term resilience through regional and global cooperation. By pursuing these strategies, SIDS can amplify their voices, secure the resources they need, and foster collaborative solutions to the climate crisis.

One of the most critical pathways for strengthening climate diplomacy for SIDS is the enhancement of international partnerships and funding mechanisms. SIDS rely heavily on external support to finance climate adaptation and mitigation projects, as their limited financial resources and small economies make it difficult to fund these initiatives independently. Strengthening partnerships with developed countries, international organizations, and non-state actors can help SIDS access the financial and technical resources they need to build resilience and achieve sustainable development.

A key area of focus is improving access to climate finance, particularly through mechanisms such as the Green Climate Fund (GCF), the Global Environment Facility (GEF), and the Adaptation Fund. These funds are designed to support developing countries in addressing climate change, but SIDS often face challenges in accessing them due to complex application processes and limited administrative capacity (Betzold & Weiler, 2017). Simplifying these processes and providing targeted technical assistance to SIDS can help them overcome these barriers and secure the funding they need. For example, the GCF's Readiness Programme has been instrumental in helping SIDS build their capacity to develop and implement climate projects (Green Climate Fund, 2021).

In addition to multilateral funds, SIDS can benefit from bilateral partnerships with developed countries. Many developed nations have established climate finance initiatives specifically targeted at SIDS, such as the Pacific Climate Change Support Programme funded by the European Union (EU, 2020). These partnerships can provide not only financial resources but also technical expertise and knowledge sharing, which are critical for building local capacity and ensuring the sustainability of climate projects. Moreover, SIDS can leverage these partnerships to advocate for increased climate finance commitments from developed countries, particularly in light of the unmet pledge to mobilize $100 billion annually by 2020 to support developing nations (Weikmans & Roberts, 2019).

Non-state actors, including civil society organizations, the private sector, and philanthropic foundations, also play a crucial role in supporting SIDS' climate efforts. For instance, the Caribbean Climate-Smart Accelerator is a public-private partnership that aims to mobilize investment in climate-resilient infrastructure and renewable energy projects in the Caribbean region (CCSA, 2021). Similarly, philanthropic organizations such as the Rockefeller Foundation have supported SIDS in developing innovative solutions to climate challenges, such as community-based adaptation projects and disaster risk reduction initiatives (Rockefeller Foundation, 2020). By fostering partnerships with these actors, SIDS can tap into additional resources and expertise to strengthen their climate diplomacy.

Another important pathway for strengthening climate diplomacy for SIDS is the use of technology and data to support evidence-based advocacy. SIDS face significant data gaps and limited access to advanced technologies, which can hinder their ability to assess climate risks, develop effective adaptation strategies, and advocate for their needs in international forums. Investing in technology and data infrastructure can help SIDS overcome these challenges and enhance their capacity to engage in climate diplomacy. Remote sensing and geographic information systems (GIS) are powerful tools for monitoring and analyzing climate-related risks, such as sea-level rise, coastal erosion, and extreme weather events. For example, the Pacific Community (SPC) has developed a GIS-based platform to support SIDS in the Pacific region in mapping climate vulnerabilities and planning adaptation measures (SPC, 2020).

Similarly, satellite imagery and drone technology can provide high-resolution data on land use, vegetation cover, and coastal changes, enabling SIDS to make informed decisions about resource management and disaster preparedness (Kumar et al., 2018). Data-driven advocacy is another critical area where technology can play a transformative role. By collecting and analyzing data on the impacts of climate change, SIDS can build a compelling evidence base to support their calls for stronger climate action and increased financial support. For instance, the Caribbean Community Climate Change Centre (CCCCC) has developed a regional climate modeling framework to project future climate scenarios and inform policy decisions (CCCCC, 2021). These models can be used to demonstrate the urgency of addressing climate change and the potential benefits of investing in adaptation and mitigation measures. In addition to data collection and analysis, digital platforms and social media can amplify SIDS' advocacy efforts by raising awareness of their challenges and mobilizing global support. For example, the #1point5tostayalive campaign, launched by Caribbean SIDS, used social media to highlight the importance of limiting global warming to 1.5 degrees Celsius and advocate for stronger climate commitments (CARICOM, 2015). Similarly, virtual conferences and webinars have enabled SIDS to participate in international climate discussions despite travel restrictions and resource constraints, ensuring that their voices are heard in global forums (UNFCCC, 2021).

Building long-term resilience to climate change requires a collaborative approach that leverages regional and global cooperation. SIDS face many common challenges, such as rising sea levels, extreme weather events, and biodiversity loss, which can be addressed more effectively through collective action. Regional organizations and initiatives play a crucial role in facilitating cooperation among SIDS and coordinating their efforts to address shared vulnerabilities. For example, the Pacific Islands Forum (PIF) has been instrumental in advancing climate action in the Pacific region through initiatives such as the Pacific Resilience Partnership and the Framework for Resilient Development in the Pacific (FRDP) (PIF, 2018). These initiatives provide a platform for SIDS to share knowledge, develop joint strategies, and advocate for their interests in international forums. Similarly, the Caribbean Community (CARICOM) has established the Caribbean Climate Change Centre (CCCCC) to support member states in implementing climate adaptation and mitigation projects and accessing climate finance (CCCCC, 2021). Global cooperation is also essential for addressing the transboundary nature of climate change and ensuring that SIDS' concerns are integrated into international climate policies. SIDS have been active participants in global climate negotiations, advocating for stronger commitments to reduce greenhouse gas emissions, increase climate finance, and address loss and damage. Their efforts have been supported by alliances with other vulnerable countries, such as the least developed countries (LDCs) and the Climate Vulnerable Forum (CVF), which amplify their collective voice and influence in international forums (Khan et al., 2020). In addition to formal negotiations, SIDS can benefit from global initiatives that promote knowledge sharing and capacity building. For example, the United Nations Development Programme (UNDP) has supported SIDS in developing National Adaptation Plans (NAPs) and implementing climate-resilient projects through its Climate Change Adaptation Portal (UNDP, 2021). Similarly, the World Bank's Small Island States Resilience Initiative (SISRI) provides technical and financial support to SIDS in building resilience to climate change and natural disasters (World Bank, 2020).

Strengthening climate diplomacy for SIDS requires a multifaceted approach that addresses their unique vulnerabilities and leverages their collective strengths. Enhancing international partnerships and funding mechanisms, leveraging technology and data for evidence-based advocacy, and building long-term resilience through regional and global cooperation are critical pathways for achieving this goal. By pursuing these strategies, SIDS can amplify their voices, secure the resources they need, and foster collaborative solutions to the climate crisis. The international community must also play its part by providing the financial, technical, and political support necessary to ensure that SIDS can adapt to the impacts of climate change and achieve sustainable development.

3.6. Conclusion

SIDS stand at the center of the global climate crisis, grappling with existential threats from rising sea levels and ecological degradation despite contributing negligible amounts to greenhouse gas emissions. This chapter has demonstrated the critical role of climate diplomacy as both a tool and a lifeline for SIDS, which enables them to amplify their voices on the international stage and advocate for urgent climate action. Through coalitions such as AOSIS, SIDS have time and time again contributed significantly to shaping global climate policy, which includes its notable achievement of securing the 1.5 degrees Celsius target embedded in the Paris Agreement. Their moral authority, rooted in disproportionate vulnerability and decades of climate injustice, has propelled them to become pivotal actors in demanding accountability from industrialized countries contributing largely to global emissions.

It is evident that SIDS face formidable challenges in sustaining these diplomatic efforts. Limited financial and technical resources, coupled with the tension between immediate development needs and long-term climate resilience plans, constrain their capacity for effective negotiation. Moreover, the slow pace of international climate finance disbursement and the bureaucratic hurdles of fund accessibility to SIDS, which already have weak administrative capacities, further exacerbate these struggles. Notwithstanding, there are several pathways for strengthening SIDS' climate diplomacy and its role in the global climate architecture. Enhancing partnerships with multilateral institutions, utilizing technology for data-driven advocacy, and fostering regional and international cooperation may empower SIDS in overcoming structural and political disadvantages. Initiatives such as the GCF's Readiness Programme and regional frameworks such as the Pacific Resilience Partnership exemplify the potential of collaborative approaches to build and secure sustainable development in small island developing nations.

Notwithstanding, the survival of SIDS is largely dependent on the international community's commitment to prioritizing equitable climate actions. Developed nations must use the Common But Differentiated Responsibility (CBDR) to honor their financial pledges, prioritize the needs of the most vulnerable nations, and take serious action to reduce their emissions to avert irreversible climate change. Simultaneously, the innovative strategies employed by SIDS, such as storytelling and legal advocacy, must be supported at a global level. It must be recognized that SIDS' fight for survival encompasses more than a regional issue; it serves as a litmus test for global solidarity and justice. In this regard, ensuring a livable future for SIDS demands a transformative shift in how the international community addresses climate change, in a manner that is swift and actionable, before these countries ultimately disappear from existence.

References

Alongi, D. M. (2015). The impact of climate change on mangrove forests. *Current Climate Change Reports*, *1*(1), 30–39.

Ananthavinayagan, T. V., Shaheed, A., Siddiqui, N. A., & Imran, M. (2024). International law and Maldives: Navigating geopolitics, trade, and sovereignty. *Law Preprints*. https://doi.org/10.33774/coe-2024-gfg4w-v2

Asian Development Bank. (2005). *Maldives tsunami: Impact and recovery. Joint needs assessment by World Bank-Asian Development Bank–UN system*. Global Facility for Disaster Reduction and Recovery (GFDRR), 9–10.

Betzold, C. (2010). Borrowing power to influence international negotiations: AOSIS in the climate change regime, 1990–1997. *Politics*, *30*(3), 155–168.

Betzold, C., & Weiler, F. (2017). Allocation of aid for adaptation to climate change: Do vulnerable countries receive more support? *International Environmental Agreements: Politics, Law and Economics*, *17*(1), 17–36.

Briguglio, L. (2016). Exposure to external shocks and economic resilience of countries: Evidence from global indicators. *Journal of Economic Studies*, *43*(6), 1057–1078.

Caribbean Climate-Smart Accelerator (CCSA). (2021). *About us*. Retrieved from https://www.caribbeanaccelerator.org

Caribbean Community Climate Change Centre (CCCCC). (2021). *Regional climate modeling framework*. Retrieved from https://www.caribbeanclimate.bz

Connell, J. (2013). *Islands at risk: Environments, economies, and contemporary change*. Edward Elgar Publishing.

Deitelhoff, N., & Wallbott, L. (2012). Beyond soft balancing: Small states and coalition-building in the ICC and climate negotiations. *Cambridge Review of International Affairs*, *25*(3), 345–366.

Depledge, J. (2010). The road less traveled: Difficulties in moving between annexes in the climate change regime. *Climate Policy*, *10*(6), 677–689.

Dornan, M. (2014). Renewable energy development in small island developing states: Challenges and opportunities. *Energy Policy*, *72*, 234–242.

European Union (EU). (2020). *Pacific climate change support programme*. Retrieved from https://ec.europa.eu

Farbotko, C. (2021). Tuvalu's climate diplomacy: Navigating legal and political challenges. *Global Environmental Politics, 21*(2), 45–62.

Farbotko, C., & Lazrus, H. (2012). The first climate refugees? Contesting global narratives of climate change in Tuvalu. *Global Environmental Change, 22*(2), 382–390.

Green, A., & Guilfoyle, D. (2024). The Australia-Tuvalu Falepili Union Treaty: Statehood and security in the face of anthropogenic climate change. *American Journal of International Law, 118*(4), 684–703.

Green Climate Fund (GCF). (2021). *Readiness programme.* Retrieved from https://www.greenclimate.fund

Hoegh-Guldberg, O., Poloczanska, E. S., Skirving, W., & Dove, S. (2017). Coral reef ecosystems under climate change and ocean acidification. *Frontiers in Marine Science, 4*, 158.

Intergovernmental Panel on Climate Change (IPCC). (2014). Small Islands. In *Climate Change 2014 – Impacts, Adaptation and Vulnerability: Part B: Regional Aspects: Working Group II Contribution to the IPCC Fifth Assessment Report* (pp. 1613–1654). chapter, Cambridge: Cambridge University Press. https://doi.org/10.1017/CBO9781107415386.009

Intergovernmental Panel on Climate Change (IPCC). (2022). Global Warming of 1.5°C: IPCC Special Report on *Impacts of Global Warming of 1.5°C above Pre-industrial Levels in Context of Strengthening Response to Climate Change, Sustainable Development, and Efforts to Eradicate Poverty.* Cambridge: Cambridge University Press. https://doi.org/10.1017/9781009157940

Intergovernmental Panel on Climate Change (IPCC). (2023). *Climate Change 2021 – The Physical Science Basis*: Working Group I Contribution to the Sixth Assessment Report of the Intergovernmental Panel on Climate Change. Cambridge: Cambridge University Press. https://doi.org/10.1017/9781009157896

Khan, M. R., Roberts, J. T., & Huq, S. (2020). The role of small island developing states in global climate negotiations. *Climate Policy, 20*(6), 720–734.

Kelman, I. (2014). No change from climate change: Vulnerability and small island developing states. *The Geographical Journal, 180*(2), 120–129.

Kumar, L., Jayasinghe, S., Gopalakrishnan, T., & Nunn, P. D. (2018). Remote sensing and GIS for coastal ecosystem assessment and management. *Remote Sensing, 10*(12), 1927.

McAdam, J. (2012). *Climate change, forced migration, and international law.* Oxford University Press.

McMichael, A. J., Woodruff, R. E., & Hales, S. (2006). Climate change and human health: Risks and responses. *The Lancet, 367*(9513), 859–869.

Mercer, J., Dominey-Howes, D., Kelman, I., & Lloyd, K. (2010). The potential for combining indigenous and Western knowledge in reducing vulnerability to environmental hazards in small island developing states. *Environmental Hazards, 9*(3), 245–256.

Mimura, N., Nurse, L., McLean, R. F., Agard, J., Briguglio, L. P., & Payet, R. (2007). Small islands. In M. L. Parry, O. F. Canziani, J. P. Palutikof, P. J. van der Linden, & C. E. Hanson (Eds.), *Climate change 2007: Impacts, adaptation and vulnerability. Contribution of Working Group II to the Fourth Assessment Report of the Intergovernmental Panel on Climate Change* (pp. 687–716). Cambridge University Press.

Mycoo, M. (2018). Beyond 1.5°C: Vulnerabilities and adaptation strategies for Caribbean small island developing states. *Regional Environmental Change, 18*(8), 2341–2353.

Nelson, V., Meadows, K., Cannon, T., Morton, J., & Martin, A. (2002). Uncertain predictions, invisible impacts, and the need to mainstream gender in climate change adaptations. *Gender & Development, 10*(2), 51–59.

Nunn, P.D., Mulgrew, K., Scott-Parker, B. et al. (2016). Spirituality and attitudes towards Nature in the Pacific Islands: insights for enabling climate-change adaptation. *Climatic Change* 136, 477–493. https://doi.org/10.1007/s10584-016-1646-9

Nurse, L. A., McLean, R. F., Agard, J., Briguglio, L. P., Duvat-Magnan, V., Pelesikoti, N., ... & Webb, A. (2014). Small islands. In *Climate change 2014: Impacts, adaptation, and vulnerability. Part B: Regional aspects. Contribution of Working Group II to the Fifth Assessment Report of the Intergovernmental Panel on Climate Change* 1613–1654. Cambridge University Press.

Pacific Resilience Partnership. (2020). *Framework for Resilient Development in the Pacific: Monitoring and Evaluation Strategy Pacific Resilience Partnership,* 1–13. https://www.spc.int/sites/default/files/tenderfiles/2023-06/1.%20Framework%20for%20Resilient%20Development%20in%20the%20Pacific.pdf

Pelling, M., & Uitto, J. I. (2001). Small island developing states: Natural disaster vulnerability and global change. *Environmental Hazards, 3*(2), 49–62.

Randall, R. (2011). Climate change and the rise of the Pacific. *The Journal of Pacific History, 46*(2), 157–169.

Region's Climate Change concerns captured in song. (n.d.). CARICOM. Retrieved August 22, 2025, from https://caricom.org/regions-climate-change-concerns-captured-in-song/

Robinson, S. A. (2020). Climate change adaptation in SIDS: A systematic review of the literature pre and post the IPCC 1.5°C report. *WIREs Climate Change, 11*(4), e653.

Rockefeller Foundation. (2020). *Building climate resilience in small island developing states.* Retrieved from https://www.rockefellerfoundation.org

Scott, D., Hall, C. M., & Gössling, S. (2012). *Tourism and climate change: Impacts, adaptation, and mitigation.* Routledge.

Secretariat of the Pacific Community (SPC). (2020). *GIS-based climate vulnerability mapping.* Retrieved from https://www.spc.int

Thomas, A., Baptiste, A., Martyr-Koller, R., Pringle, P., & Rhiney, K. (2020). Climate change and small island developing states. *Annual Review of Environment and Resources, 45*, 1–27.

United Nations Development Programme (UNDP). (2021). *Climate change adaptation portal.* Retrieved from https://www.adaptation-undp.org

United Nations Framework Convention on Climate Change (UNFCCC). (2021). *Virtual climate conferences and webinars.* Retrieved from https://unfccc.int

Vanhala, L., & Hestbaek, C. (2016). Framing climate change loss and damage in UNFCCC negotiations. *Global Environmental Politics, 16*(4), 111–129.

Weikmans, R., & Roberts, J. T. (2019). The international climate finance accounting muddle: Is there hope on the horizon? *Climate and Development, 11*(2), 97–111.

World Bank. (2018). *Dominica: Climate resilience and recovery plan.* World Bank Group.

World Bank. (2020). *Small Island States Resilience Initiative (SISRI).* Retrieved from https://www.worldbank.org

4

Guardians of the Future

How Women Hold the Key to Just Climate Transitions

Vesselin Popovski, Rajat Shandilya, Shayani Sarkar

4.1 Introduction

Women have made significant contribution toward the betterment of society, and the field of climate change and sustainability is one of them. It has been in the early twenty-first century that the efforts of women in this area began to be acknowledged and appreciated. Feedback from women and the issues faced by them have been described as "imperative" by the United Nations (United Nations Entity for Gender Equality and the Empowerment of Women, 2015) and "critical" by the Population Reference Bureau (Yavinsky, 2012). A report by the World Health Organization concluded that "incorporating gender-based analysis would "provide more effective climate change mitigation and adaptation" (World Health Organization, 2015).

However, the escalating climate crisis has underscored the urgent need for inclusive and transformative policies to address its far-reaching impacts. It is a fact that the consequences of climate change are not perceived uniformly, and that climate change has a greater impact on women than on men.

The world has reached a stage wherein the only way forward to deal with climate change is by means of unparalleled joint global action. In addition to exploring the multifaceted roles that women already play in the development and implementation of climate change policies, emphasizing their contributions as leaders, advocates, and visionaries, this chapter also argues for the further advancement of women to that of a major contributor to climate diplomacy to amplify the voice for just and sustainable climate actions. It is pertinent to note that by providing women with access to financial resources and assistance in legal capacity-building, the expertise of women (affected or otherwise) can assist in creating an impartial forum for improved collaborative deliberations on climate diplomacy, which can aid in a just climate transition.

Analyzing participation data, it has come to evidence that women are minimally present in any breakthrough discussions or deal. Their concerns are largely omitted in major environmental deals by governments and international organizations; consequently, as a result, the design of adaptation and mitigation efforts tailored at these levels targets the remedy to climate change as a gender-neutral problem when it is not (Singh, 2017). It is a harsh fact that even today, women are unfortunately underrepresented in several fields related to the study of climate change. An analysis of US atmospheric science doctoral programs reveals that women comprised 17 percent of tenure-track and tenured faculty, with even smaller proportions at higher ranks, and 53 percent of departments had two or fewer women faculty (MacPhee & Canetto, 2015).

It is argued that when women are overlooked as scholars and decision makers, the world fails to take advantage of its full human capacity, which is needed for issues as urgent as climate change (Wyer et al., 2013). Women may also take more collaborative approaches, especially in negotiations, and may pay more attention to disadvantaged groups and to the natural environment (Blanchard, 2003). Gender has become an issue because of women's essential roles in managing resources such as water, forests, and energy and as women lead fights for environmental protection (Buckingham-Hatfield, 2005). A general concern has been expressed about the need to highlight the work of women and to include more women in major committees in order to provide gender balance, social justice, and inspiration to young women to enter careers in science (Buck et al., 2014). A report prepared by UN Women, the Mary Robinson Foundation—Climate Justice, the Global Gender and Climate Alliance and the UNFCCC recognizes the structural inequalities that impede the representation of women in climate science, negotiations, and policies and recommends greater gender balance in the UNFCCC and national delegations (UN Women & Mary Robinson Foundation, 2013).

Women's environmental leadership can not only bring about improvements in representational justice and an increased focus on gender-specific environmental impacts, but it can also lead to stronger and more effective environmental action (OECD & IEA, 2021) Female participation and active involvement in the decision-making process shall lead to ensuring better environmental safeguards and, in turn, also ensure the highest level of productivity and outcomes. Research supports that a higher presence of women in public sector decision-making bodies translates into more ambitious climate goals and policies. Evidence from the business sector shows that when women occupy at least 30 percent of board seats, they start making a difference in climate governance within companies (The Sasakawa Peace Foundation & Bloomberg NEF, 2020) In contemporary times, globally, women are leaving a powerful impact in the field of climate change and sustainability.

There have been multiple instances from around the world that exhibit women excelling in leadership roles when it comes to climate diplomacy. In Audre Lorde's words, "the master's tools will never dismantle the master's house." (Lorde, 2018). Women's leadership in government corresponds with greater environmental protections (Nugent & Shandra, 2009), more stringent climate policies (Mavisakalyan & Tarverdi, 2019), and the ratification of environmental treaties (Norgaard & York, 2016) as seen, for example, in increased favorable environmental outcomes in the European Parliament and US Congress (Ramstetter & Habersack, 2020). Moreover, the environmental and climate policies that do get passed when women are in leadership positions are likely to be more effective and promote equality (Cook et al., 2019). It is a high time to formally recognize and acknowledge the significant contributions made by women in advocating for climate change and also work toward bringing and ensuring "transformative climate action." In the same vein, another important point to be mentioned here is that gender-responsive climate change mitigation and adaptation need active participation and inclusion of women in the process, as power has a significant impact on the development of adaptive abilities by serving as the primary explanatory framework for gendered power relations (FRONTIERS?)[1] Women, without a doubt, have contributed immensely to climate change research and policy and to broader analyses of global environmental issues (Sachs, 2014).

4.2 Effective Contribution of Women in Climate Leadership Roles

Women have played important leadership roles in international climate policy. A few influential figures are as follows:

1. Christiana Figueres is often regarded as the torchbearer and a leading expert in the field of climate change, and she leads the international

climate negotiations as the Executive Secretary of the UN Framework Convention on Climate Change (UNFCCC).

2. Hilda Heine is regarded as having made multiple key decisions regarding climate change for the betterment of the Pacific Nations, specifically low-lying Pacific islands. The effect of this can be seen by the fact that over the last few years, many of her suggestions have also been duly acknowledged and interestingly implemented by the government. One such example in this regard is the implementation of the coastal protection system.
3. Kate Marvel is credited with discovering the relationship between global rainfall patterns and the change in the nature of fundamental human activities.
4. Sunita Narain has also been a very notable and celebrated name in the field of climate change and policy. Her niche area is related to rainwater harvesting and animal (particularly tiger) conservation. In her opinion, the voice of the minority from the south has not received attention.
5. Greta Thurnberg has become a household name when it comes to the aspect of climate change and sustainability. It was in the year 2018 that the then fifteen-year-old Greta Thunberg (TIME, 2019) started a school strike in Sweden to draw the attention of the world over the worsening global climatic conditions, inviting climate crises.

Contextually, there are multiple ways that can play an assistive role in identifying and acknowledging women who have significantly contributed to the field of climate change and sustainable development. The first parameter in the same vein is getting featured in the highly regarded list of authors published by the United Nations and various other organizations, for example, the Intergovernmental Panel on Climate Change and the UN Framework Convention on Climate Change (UNFCCC).

In the same regard, the second parameter can be seen from the point of view of academic contributions. This can be done by examining the academic credibility and academic contribution of such women. The third parameter can be said to be, in the opinion of the author, to scrutinize the point of membership in prestigious global change committees of the International Council for Science (ICSU).

Seven cities around the world have appointed women as Chief Heat Officers (CHOs) to take action against extreme heat due to climate change (Talib, 2022).

In February 2024, the EIB Group established the Women Climate Leaders Network, bringing together forty-seven women leaders from the commercial sector from all twenty-seven EU Member States to collaborate on innovative business models and strategies to facilitate the low-carbon and green transition (European Investment Bank, 2024). Thus, based on the above segmental discussion, it is safe to say that categorizing women simply as a vulnerable group without acknowledging their role in climate change responses serves to emphasize women's needs while their leadership and active participation are unsupported and not quantified.[2] In this article, the author has made an attempt to critically examine the current status of women's leadership. The fundamental line of argument will advocate for the fact that, in today's contemporary scenario where climate change has become a pressing issue, women have a crucial role to play in the same regard.

4.3 Role of Indigenous Women

Multiple Indigenous communities have relied more on giving environmental knowledge to the next generations. This also strengthens the base with regard to historical knowledge and makes people aware of various strategies that should be adopted with respect to more sustainable practices. There is an increasing push to understand how genders are differently affected by climate change and to use the various perspectives that are not always seen to initiate change. Overall, there is a general push to involve more genders in larger climate issues to see how best to address these issues (United Nations, 2009). One such practical example is Hindou Oumarou Ibrahim, a renowned name as an Indigenous activist who is based in Chad. She is actively working in Chad to spread awareness about climate change. She is working to educate people on climate change and the impacts it is having on the Earth. She aims to create a positive environment for women to work within the climate movement (Staff, 2019). Katharine Wilkinson has worked on a book called the All We Can Save Project that brings together the voices of sixty women who are doing work on climate change on the ground today. She aims to include more voices of women in order to open up the climate change discussion to different perspectives as well as show the readers the ways we can help to combat climate change together.[3]

4.4 Gendered Lens on Climate Action: Feminist Perspectives Within SDG 13 and the Paris Agreement

Under the ambit of International Law, the highly influential and important Paris Agreement on Climate Change (Paris Agreement), in the opinion of the researcher, can be regarded as a "hybrid treaty" as the Paris Agreement is a combination of both hard law and soft law in one place and an example of this is the flexibility that the Paris Agreement gives various states regarding the decision of their self-ascertained nationally determined commitments (NDCs). In the same vein, after the decision of their NDCs voluntarily, these commitments convert to have a binding effect.

The discipline of international law has gradually evolved by following a top-down approach wherein the concerned bodies which overlook and regulate international law and governance on a global level used to impose sanctions on a particular country that does not follow the international obligations.

However, when it comes to the aspect of the Paris Agreement, an important thing to notice here is that the Paris Agreement is conventionally and principally different from the multiple bodies and fundamental principles of international law, as it gives more importance on collaborative international effort. This is evident in the fact that the Paris Agreement has established a facilitation committee in place of a sanction committee in order to ensure close harmony and alignment between the binding obligations toward climate change and its related aspects, as well as the improvement and implementation of various policies while ensuring fair access, transparency, and accountability. Based on the above line of arguments, it is safe to say that an approach of this nature toward ensuring the optimum level of global governance is more in line with the long-term goals in the agenda, which can primarily be said to be normative in nature. Such a flexible approach provides a great amount of individual autonomy to a state in general. On practical grounds, it can be said that the liberty to set one's own target based on the internal circumstances of the nation can also turn out to be of immense importance to the concerned nation.

Christina Figueres from Costa Rica was the leader of the negotiations leading to the Paris Agreement, who demonstrated excellent female leadership. At this juncture, it is essential to discuss the South Asian feminist angle to climate diplomacy. India has been making strides in bridging the gender gap in shaping India's climate policies, sustainability initiatives, as well as advocacy efforts. Specifically, the Bagepalli Clean Development Mechanism (CDM) Biogas Project recognizes the crucial role played by women in household energy management and the health advantages of cleaner cooking fuels for women and children (Subbarao & Lloyd, 2011). They were actively involved in the decision-making process, influencing the shift from traditional biomass fuels to cleaner biogas alternatives, thereby helping to strengthen community acceptance of the project and its long-term success.

The 2008 National Action Plan on Climate Change (NAPCC) formulated by the Government of India fails to recognize the gender dimensions of climate change (CDKN, 2016). Though it recognizes the impacts of climate change on society, it hardly incorporates any gender-specific measures (Parikh, 2012). However, climate policy planning incorporated by numerous Indian states in many pilot projects has made special gender considerations. For example, in Madhya Pradesh, the State Action Plan on Climate Change (SAPCC) promoting climate-resilient agriculture has advocated for gender considerations. Uttar Pradesh has integrated gender sensitivity into its SAPCC. It specifically discusses the links between gender and climate adaptation and mitigation, which will positively help in fostering more inclusive climate-compatible development processes in the state (CDKN, 2016). Uttarakhand, West Bengal, and Bihar, to name a few, have also advocated for gender considerations in climate-resilient agricultural practices and have included various sections on gender and environmental budgets in climate policy planning.

Yet another example is of Godavari Dange, a farm entrepreneur from Osmanabad in India, who, following the severe droughts between 2007–8, led the climate-resilient "One Acre Farming Model" where female farmers have successfully battled drought and farmer suicides by adopting low-input sustainable farming techniques, including efficient water usage, organic farming, mixed cropping, and increased crop cycles. (Reference no 10)A successful outcome was that while all of these examples demonstrate women's leadership and ability to cope with climate change at the local level, these actions continue to remain insufficient as more efforts are needed to boost women's representation. As highlighted in the Gender Action Plan that was adopted at COP23, a just transition to sustainable development can only be achieved if women are actively involved in the development and implementation of all aspects surrounding climate change mitigation and adaption (UN Women, 2020). However, in many countries, women continue to be poorly represented in national parliaments, governments, and even local decision-making bodies.

Incorporating feminist perspectives into climate change policies can play a significant role in enhancing their effectiveness and fairness. This is also because no area of policy is gender-neutral, and hence integrating gender democracy into climate policy is extremely essential. In India, Dr. Vandana Shiva's name stands out in the arena of environmental sustainability and its intersection with feminism. Specifically, her work links ecofeminism, environmental issues, women's rights, and social justice. Her work starts from the Chipko movement which was a historical milestone in the field of women and environmental sustainability.

Her work elaborates on the role of social and political structures, patriarchy, and capitalism in affecting resources and environmental structures. Patriarchal structures contribute to environmental degradation and the marginalization of women in rural and urban communities. The concept of intersectional feminism can bring together the struggles of lower-income rural women (UN Women, 2020). However, in this context, it is also pertinent to note that embracing a feminist perspective entails more than just focusing on women. It also signifies establishing a more comprehensive and inclusive awareness of climate change along with emphasizing the interconnected nature of its effects on every aspect of life. It primarily advocates for equitable and just climate action. When we talk about intersectional feminism in the context of climate change, we need to focus on the experiences of those women who have experienced climate change.

A gendered lens in climate policies in India would require persistent policy reforms, mainstreaming gender into climate policies, raising women's voices in decision-making, and addressing gender-specific climate impacts. More women are needed in climate leadership and finance because, even in 2024, the vast majority of women in emerging economies do not have adequate access to cover climate-related losses emerging from floods, droughts, heatwaves, and soil erosion.

In addition to India, Bangladesh has also been making significant efforts to incorporate climate legislation. The 15th Constitutional Amendment of Bangladesh prioritizes environmental and climate protection. A study recommends engaging non-governmental organizations and adaptation programs to pull climate change funds. It suggests track two diplomacy and proper training for diplomats. Additionally, it suggests proper plans for a climate fund and arrangements for an international summit on climate change, as well as appointing a special envoy for climate change (Nomani & Hossain, 2022).

In South Asia, minimal female participation is present in climate policymaking, despite the fact that a large percentage of the population depends on agriculture. Though women tend to be disproportionately affected by climate change, especially in developing nations (Yavinsky, 2012), gender as a concept in itself has been missing from the debate on climate change. It is specifically highlighted that addressing climate change through a gendered lens is not just a question of fairness but also a fundamental requirement. Engaging women in climate governance, policymaking, and on-the-ground activities can result in equitable and sustainable solutions, ensuring that most of the population's challenges and vulnerabilities are successfully dealt with in the global battle against the impacts of climate change.

4.5 Women and Climate Change: Shifting Narratives and Emerging Leadership

The narrative of women as mere victims of climate change has begun to shift. Climate finance mechanisms adopted by the Green Climate Fund and similar frameworks ensure that climate change projects involve women in project selection, assessment, as well as in addressing gender concerns (Asian Development Bank, 2014).

Women and girls in Central Africa play a central role in building community resilience. They have emerged as leaders in developing innovative solutions such as drought resistance and better soil management. However, despite women accounting for 70 percent of food production in Africa, less than 20 percent of the land is owned by women (World Economic Forum, 2018). This lack of ownership, coupled with limited decision-making opportunities, excludes women from significant roles in climate diplomacy. Further research has shown that if women farmers are given the same opportunities as male farmers, farm yields will increase by 20–30 percent, and hunger will reduce by 17 percent.

The existing literature emphasizes the significance of involving indigenous peoples in climate diplomacy in order to preserve biodiversity. This is specifically important for indigenous women, as their interconnectedness with the environment often reflects cultural and ecological identities. Considering the relationship that exists between indigenous peoples and the environment, the non-existence of natural resources could also mean deprivation of the right to life, food security issues, and health problems, such as lack of pollution-free air and water. The loss of natural resources can create a sense of deprivation of essential rights for indigenous populations. Furthermore, ecological destruction affects not just their subsistence but also their cultural and shared identities, emphasizing the importance of an inclusive climate strategy that highlights indigenous perspectives, particularly those of women.

The status quo is such that men are physiologically better equipped to withstand the physical impact of a disaster than women (Neumayer & Plümper, 2007). For example, the 2005 Oxfam International report on the December 2004 tsunami in South and Southeast Asia demonstrates that the differences in self-rescue abilities were partly determined by a lack of learned survival skills and not merely physiological differences. It was also stated that women have different propensities to die from diseases exacerbated by disasters (Oxfam International, 2005).

Specifically in India, female farmers face increased challenges due to climate variability, affecting food security and livelihoods. During water shortages, women's participation in the process involving the collection of water aggravates burdens as well as health hazards. Large-scale natural disasters worsen these problems by damaging the existing social order and limiting the availability of food, sanitation, and medical facilities.

Social norms constitute one of the major reasons for gendered disaster vulnerability. For instance, the 2004 tsunami that hit the coast of Indonesia saw many women in rural coastal areas at home while men were out at sea fishing or away from home. Similarly, in India, many women were waiting at the seashore for fishermen to arrive. In both cases, more men survived because the waves only gather height and strength as they approach the shore.[4] A study on the 2001 Hurricane 07B in the Godavari Delta found that migrant, low-caste women—often landless agricultural labourers—constitute the most vulnerable group (Neumayer & Plümper, 2007). Further, it has been documented that law and order are difficult to sustain during conflicts, and female victims often have to seek refuge in makeshift camps where girls and women are at a heightened risk of sexual abuse or violence.

Another notable factor is that pertaining to the gender gap in life expectancy during disasters, which highlights systemic inequalities. While men and women face distinct challenges, boosting women's social and economic standing has the potential to undo these disparities. Research also shows that women generally live more sustainably than men and contribute less to climate change (Johnsson-Latham, 2007).

Efforts such as the Paris Agreement's gender-responsive policies are admirable, but their non-binding status limits their effectiveness. It is a fact that if there is no gender representation in climate policies, the outcome will not be an inclusive or gendered approach. In this specific context, representation constitutes a critical issue. Anne Philips' "Politics of Presence" highlights the importance of descriptive and substantive representation. Descriptive representation relates to the number of women in policymaking roles, while substantive representation specifically highlights their ability to influence policy outcomes. For instance, the European Grean Deal is completely gender-blind despite using terms like "just and inclusive." (Phillips, 1995). Impact assessment could be one particular step in viewing climate diplomacy through a gendered lens. However, gender is often ignored in these assessments. Another pertinent aspect of gender mainstreaming is that gender equality should not only be perceived as equality between men and women, but also as that which addresses heterogeneity within these categories.

The Gender Equality Strategy discusses some ways in which climate change is gendered but provides no specific details. Studies on gender mainstreaming—and in this case, climate change mainstreaming—show that organizational structures are not designed for cooperation, coordination, and joint decision-making. Hence, unless gender inequality is addressed, women leaders in climate diplomacy will continue to be underrepresented compared to men. The goal of gender-based leadership in climate diplomacy can only be successful if development and climate change actions are equitable and place women's empowerment and resolution of gender inequality at the center.

Sustainable climate change interventions should include a gender-based approach to development. On the topic of sustainable development, the gender-based approach addresses both practical and strategic gender needs. When discussing climate change, institutional frameworks for gender-sensitive social relations and implementation of climate policies need to be created. Awareness programs must be developed to train women in gender discourse, and disaster management initiatives must include women in leadership roles. Documenting the knowledge and experiences, particularly of rural women, can enhance policy-making and promote gender-responsive mitigation mechanisms. Furthermore, countries should be encouraged to promote better attitudes and create policies that include women, specifically in leadership roles.

The resolution of gender inequality is crucial for achieving equitable and long-term climate solutions. The role of women in climate diplomacy demands deliberate initiatives to incorporate their opinions, abilities, and experiences. Gender-responsive mechanisms, empowering women, and guaranteeing their presence in decision-making processes can help the international community build more equitable and productive solutions to climate concerns.

4.6 Women Leading the Fight Against Climate Change and the Way Forward

Women have played pivotal roles in leadership and diplomacy, particularly in addressing climate change. However, their representation in climate diplomacy remains disproportionately low. When we talk about women in leadership positions, research states that in the COP framework of the United Nations, only 18 percent of the delegates sent to conventions were women. Some prominent women leaders include Dame Meg Taylor, former secretary general of the Pacific Islands Forum Secretariat, who is popularly known for her work in diplomacy and climate change negotiations. Kathy Jetnil-Kijiner is yet another prominent woman who is well-known for her poetry that deals with climate change and other environmental issues (Carter & Howard, 2021). In 1990, the establishment of the Alliance of Small Island States (AOSIS) witnessed leaders like Nauru Ambassador Marlene Moses using their positions to elevate climate change, sustainable development, and the needs of the Pacific nations.

A close look into the statistical representation of women in climate diplomacy at COP 1995 shows that 18 percent of the delegates were women (Carter & Howard, 2021). Between 2008 and 2012, 32 percent of the United Nations Framework Convention on Climate Change (UNFCCC) delegates were women (Höne, 2013). COP23 saw 38 percent women (Craft & McCraine, 2019). However, at the COP Summit in 2019, 80 percent of the heads of delegation were men. Regarding diplomacy, research further states that "globally only 26 women serve as heads of government and state" (UN Women, 2024). Data as of Jan 11, 2024, shows that 23 percent of ministerial positions are held by women, and in 141 countries, women make less than a third of cabinet ministers. Furthermore, in many countries there has been no female representation in the cabinet.

Gender-specific references were incorporated into the preamble, purpose, adaptation, finance, and capacity-building measures in COP21 in 2015 (Columbia University's Energy Policy and Climate Program, 2022). "COP Decision 1/CP 21," which annexes the Paris Agreement, mentions gender equality and empowerment of women as fundamental to climate action (Columbia University's Energy Policy and Climate Program, 2022). Through the Women Watch factsheet on climate change, the United Nations also recommended that governments incorporate a gender perspective into their national policies, action plans, and other measures to incorporate sustainable development and climate (Columbia University's Energy Policy and Climate Program, 2022). It further states that China, the European Union, and the United States, as influential geopolitical actors in climate negotiations, should take a bold stand on promoting women's leadership in climate tasks by specifically ensuring equal representation in climate change bodies. The United States included women's representation in disaster response, health, and climate adaptation. Another significant policy recommendation is that delegates attending UNFCCC meetings should also be trained on gender equality and climate change. Further emphasis needs to be put on the representation of indigenous women. Indigenous women have been at the forefront when it comes to leading women against natural resource exploitation and climate change. For example, the voices and perspectives of indigenous women on the front line of the fight against climate change can sharpen the policy proposals of the United States and other proposals and widen their impact (Columbia University's Energy Policy and Climate Program, 2022). The paper further recommended including young women in policy forums as well as in international negotiations. Furthermore, to ensure that accessibility is not an issue, women should be given travel support and assistance to support the mission of achieving 50 percent participation in climate-related activities (Columbia University's Energy Policy and Climate Program, 2022).

The Summit of the Future (SOTF) is a relatively recent phenomenon in the arena of climate diplomacy. Numerous provisions have been inculcated, focusing on the empowerment of women and girls. Article 8 of the Pact for the Future talks about gender equality and empowerment. It further talks about the achievement of full potential and sustainable development, which is impossible if women and girls are denied basic human rights and opportunities. It removes all legal, social, and economic barriers to achieve gender equality. It provides for targeted and accelerated action to eradicate violence and harassment against women and girls. It also refers to the gender gap as the linkage between poverty and gender equality. It further grants women rights as well as access to resources, ownership, and control over land, property, and financial services. Universal access to reproductive health and reproductive rights is also emphasized. Furthermore, while Article 15 specifically acknowledges that goals cannot be achieved without the full, equal, and meaningful participation and representation of women in political and economic life, Article 19 recognizes the role of women as agents of peace. It also emphasizes the need to redouble efforts to achieve gender equality. Article 31 acknowledges the fact that science and technology can improve women's lives. It also further talks about gender-related risks, including sexual and gender-based violence, and trafficking. Article 36 talks about youth participating in climate diplomacy, and Article 38 provides for inclusive governance (United Nations, 2024).

The overrepresentation of men in climate diplomacy is a serious problem and can be detrimental to the overall results of climate diplomacy. Given the knowledge and experiences held by women, female representation can improve social outcomes in relation to the environment and peace-building efforts. However, at present, there is a dearth of female leaders in climate activism, which could arise due to issues pertaining to weather, transport, and distance. When we talk about women's representation in climate change, we discuss two specific aspects: the first is the presence of women at the negotiation table, and the second is the need for a gendered approach to climate policies. It is absolutely necessary that we work on structural ideas because, if not, then decision-making spaces could be completely closed off to women.

It is essential that gender parity in climate leadership is achieved by 2068 (The Guardian, 2021). Female representation in national parliaments across ninety-one countries correlates with more stringent climate change policies and lower carbon emissions. This is a testament to the fact that gender equality can improve societal outcomes in relation to the environment and peace-building efforts (The Guardian, 2021).

There are numerous inspirational women who are leading the fight against climate change. Christina Figueres is a Costa Rican diplomat who served as the Executive Secretary of the UNFCCC. She played a significant role in injecting a unique sense of optimism and further successfully steered the world leaders to reach the Paris Agreement in 2015. She has actively shed light on matters of importance like gender diplomacy and climate change. Another such personality is Rhiana Gunn-Wright, who currently serves as the Climate Policy Director at the Roosevelt Institute, headquartered in New York City. She has actively worked on bringing about a holistic approach to the functioning behind the scenes at New Consensus, a think-tank championed by Alexandria Ocasio-Cortez, which aims to develop and promote the Green New Deal. The president of the Marshall Islands, Hilda Cathy Heine, is another woman who has been one of the most active individuals in the international politics of climate change. She has taken numerous important decisions in the arena of climate change in the Pacific nations, specifically that of low-lying Pacific Islands. It was under her leadership that the Marshall Islands became the first nation to submit new, binding climate targets that were in accordance with the Paris Agreement.

In addition to women holding leadership positions across international think-tanks and organizations, there are numerous inspirational women working at the grassroots level. Hindou Oumarou Ibrahim, from the Mbororo pastoralist community in Chad, is an environmental activist and geographer. She has spent the last ten years working to bridge the gap that exists between international decisions on climate change and the status quo reality at the grassroots level. She also founded the Association of Indigenous Peul Women and Peoples of Chad (AFPAT), which is a community-based organization that primarily focuses on the promotion of rights of women and girls within their Mbororo community and inspires leadership and advocacy in the arena of environmental protection. In the historic 2015 climate change meeting in Paris, she was a key leader among the indigenous groups that successfully lobbied to have their rights recognized. Kate Marvel, Senior Climate Scientist at Project Drawdown, is yet another woman in the said field. During her time as a postdoctoral researcher with the Larence Livermore National Laboratory, she discovered that human activity has changed global rainfall patterns. She also works as an Associate Scientist at NASA Goddard Institute for Space Studies at Columbia University, wherein during her study of climate change and tree rings, she discovered that climate change has been affecting droughts since 1900. Similarly, various other women, young and old, like Greta Thunberg, Ellen Page, Kotchakorn Voraakhom, Miranda Wang, Katharine Wilkinson, Changua Wu, Inger Andersen, and so on, from different walks of life and professional standing have made significant contributions to climate advocacy, which in itself is a testament to the impact of women's leadership in addressing global environmental challenges. Whether their actions have been made through grassroots-level activism, scientific and technological solutions, advocacy, or sustainable design, it is a depiction of how women can play a major role in leading the path to a sustainable future.

The representation of women in climate diplomacy and leadership is critical for developing just and sustainable climate policies, since their engagement has been proven to enhance social outcomes and environmental development. To achieve gender equality in climate leadership, focused initiatives are required, such as enabling women by means of policy changes, capacity training, as well as inclusive representation.

Notes

1 Frontiers (n.d.b).
2 Frontiers (n.d.a).
3 *The all we can save project*. The All We Can Save Project. (n.d.). https://www.allwecansave.earth/
4 http://ndl.ethernet.edu.et/bitstream/123456789/70142/1/Mohamed%20Behnassi.pdf (p. 77)

References

Asian Development Bank. (2014). *Climate finance work with women* (p. 12). Retrieved from https://www.adb.org/sites/default/files/publication/42881/climate-finance-work-women.pdf

Behnassi, M. (N.d.). *Environmental issues in developing countries* (p. 77). Retrieved from http://ndl.ethernet.edu.et/bitstream/123456789/70142/1/Mohamed%20Behnassi.pdf

Buck, H., Gammon, A. R., & Preston, C. J. (2014). Gender and geoengineering. *Hypatia*, *29*(3), 651–669. https://doi.org/10.1111/hypa.12083

Buckingham-Hatfield, S. (2005). *Gender and environment*. Routledge.

Carter, G., & Howard, E. (2021). Pacific women in climate change negotiations. *Policy Forum*. Retrieved from https://www.policyforum.net/pacific-women-in-climate-change-negotiations/

CDKN. (2016). *India gender report* (p. 9). Retrieved from https://cdkn.org/sites/default/files/files/CDKN_India-Gender_report_WEB.pdf

Columbia University's Energy Policy and Climate Program. (2022). *Women and gender in climate diplomacy*. Retrieved from https://www.energypolicy.columbia.edu/publications/women-and-gender-climate-diplomacy/

Cook, N. J., Grillos, T., & Andersson, K. P. (2019). Gender quotas increase the equality and effectiveness of climate policy interventions. *Nature Climate Change*, *9*, 330–334.

Craft, B., & McCraine, S. (2019). *Women in the UN climate negotiations: Are we tipping the balance?* International Institute for Environment and Development. Retrieved from https://www.iied.org/women-un-climate-negotiations-are-we-tipping-balance

European Investment Bank. (2024). *EIB gender equality and women's economic empowerment – Overview 2024*. Retrieved from https://www.eib.org

Frontiers. (N.d.a). *Climate change and gender in Africa: A review of impact and gender-responsive solutions*. Retrieved from

Frontiers. (N.d.b). *Climate-induced migrant's hopeful journey toward security: Pushing the boundaries of gendered vulnerability and adaptability in Bangladesh*. Frontiers.

Höne, K. (2013). The 'Doha miracle'? Where are the women in climate change negotiations? *E-International Relations*. Retrieved from https://www.e-ir.info/2013/01/18/the-doha-miracle-where-are-the-women-in-climate-change-negotiations/

Johnsson-Latham, G. (2007). *A study on gender equality as a prerequisite for sustainable development*. Report to the Environment Advisory Council, Sweden.

Lorde, A. (2018). *The master's tools will never dismantle the master's house*. Penguin UK.

MacPhee, D., & Canetto, S. S. (2015). Women in academic atmospheric sciences. *Bulletin of the American Meteorological Society*, *96*(1), 59–67. https://doi.org/10.1175/bams-d-12-00215.1

Mavisakalyan, A., & Tarverdi, Y. (2019). Gender and climate change: Do female parliamentarians make a difference? *European Journal of Political Economy*, *56*, 151–164.

Neumayer, E., & Plümper, T. (2007). The gendered nature of natural disasters: The impact of catastrophic events on the gender gap in life expectancy, 1981–2002. *Annals of the Association of American Geographers*, *97*(3), 551–566. https://doi.org/10.1111/j.1467-8306.2007.00563.x

Nomani, S., & Hossain, M. T. (2022). Bangladesh in global environmental politics: Focus on climate diplomacy. *Journal of Alternative Perspectives in the Social Sciences*, *11*(3), 415.

Norgaard, K., & York, R. (2016). Gender equality and state environmentalism. *Gender & Society*, *19*, 506–522.

Nugent, C., & Shandra, J. M. (2009). State environmental protection efforts, women's status, and world polity. *Organization & Environment*, *22*, 208–229.

OECD & IEA. (2021). *Women in senior management roles at energy firms remains stubbornly low, but efforts to improve gender diversity are moving apace*. Retrieved from https://www.iea.org

Oxfam International. (2005). *The tsunami's impact on women*. Oxfam International. Briefing Note.

Parikh, J. (2012). Climate change on human security. *Cadmus Journal*, *1*(4). Retrieved from https://www.cadmusjournal.org/files/pdfreprints/vol1issue4/Climate_Change_on_Human_Security_Jyoti_Parikh.pdf

Phillips, A. (1995). *The politics of presence*. Oxford University Press.

Ramstetter, L., & Habersack, F. (2020). Do women make a difference? Analyzing environmental attitudes and actions of members of the European Parliament. *Environmental Politics*, *29*, 1063–1084.

Sachs, C. (2014). *Women working in the environment: Resourceful natures*. Taylor & Francis.

Singh, A. (2017). Gendering climate change discourse in India. *International Research Journal of Social Sciences*, *6*(6), 32–34.
Subbarao, S., & Lloyd, B. (2011). Can the clean development mechanism (CDM) deliver? *Energy Policy*, *39*(3), 1600–1611. https://doi.org/10.1016/j.enpol.2010.12.039
The Guardian. (2021). Female leaders and the climate crisis: COP26 and the need for diversity. *The Guardian*. Retrieved from https://www.theguardian.com/commentisfree/2021/nov/11/female-leaders-climate-crisis-cop26-diverse
The Sasawaka Peace Foundation & Bloomberg NEF. (2020). *Gender diversity and climate innovation.*
Time. (2019). *Now I am speaking to the whole world: How teen climate activist Greta Thunberg got everyone to listen.*
UN Women & Mary Robinson Foundation—Climate Justice. (2013). *The full view: Advancing the goal of gender balance in multilateral and intergovernmental processes.*
UN Women. (2020). *Just transition, decent work, and cimate resilience article*. Retrieved from https://un-women.medium.com/intersectional-feminism-what-it-means-and-why-it-matters-right-now-7743bfa16757
UN Women. (2024). *In the biggest electoral year in history, 113 countries have never had a woman head of state, new UN Women data shows*. Retrieved from https://www.unwomen.org/en/news-stories/press-release/2024/06/in-the-biggest-electoral-year-in-history-113-countries-have-never-had-a-woman-head-of-state-new-un-women-data-shows
United Nations Entity for Gender Equality and the Empowerment of Women. (2015). *Women, gender equality and climate change*. Retrieved from https://www.un.org/womenwatch
United Nations. (2024). *State of the future pact for the future* (Rev. 3). Retrieved from https://www.un.org/sites/un2.un.org/files/sotf-pact-for-the-future-rev.3.pdf
United Nations. (2009). *Women...in the shadow of climate change*. United Nations. Retrieved from https://www.un.org/en/chronicle/article/womenin-shadow-climate-change
World Economic Forum. (2018). *Women farmers, food production, and land rights*. Retrieved from https://www.weforum.org/stories/2018/03/women-farmers-food-production-land-rights
World Health Organization. (2015). *Gender, climate change, and health.*
Wyer, M., Barbercheck, M., Cookmeyer, D., Ozturk, H., & Wayne, M. (2013). *Women, science, and technology: A reader in feminist science studies*. Routledge.
Yavinsky, R. W. (2012). *Women more vulnerable than men to climate change*. Population Reference Bureau.

Part II

Tools and Actors of Climate Diplomacy

5

Glocal Strategies in Climate Diplomacy

Case of Türkiye

Şuay Nilhan Açıkalın and Betül Uygur

5.1 Introduction

Climate change and its consequences are one of the hot debates in world politics. The climate problem presents significant obstacles. Human activity has warmed the atmosphere, seas, and land (Ghil & Lucarini, 2020). Since 1850, the last four decades have been warmer than the previous decades. Greenhouse gas (GHG) emissions have already raised temperatures, endangering lives and livelihoods. Extreme weather occurrences have an increasing frequency and severity (Thuiller, 2007). The consequences also create significant geopolitical concerns that go to the core of international politics.

As outlined, tackling the consequences of climate change is more controversial than the phenomenon itself, which is multifaceted and structured around multiple factors. The complex nature of climate change politics also reflects diplomacy (Sending et al., 2019). National and international agendas about climate change have found ground in diplomatic efforts in the last 50 years. Also, the climate change framework has created its own diplomatic circle, efforts, and legal achievements through national and international agendas (Sprinz & Luterbacher, 1996; Bodansky, 2001).

The notion of climate diplomacy began to be used more often in the international system regarding various levels of actors and implementation (Mabey et al., 2013). Although there are multiple definitions of climate diplomacy, its scope, aim, actors, methods, and target groups are broader than ever. Nation-states also make climate diplomacy one of the priorities of their foreign policy agenda and a vital component of soft power that develops personalized climate diplomacy tools while contributing to international collaborations (Tykhomyrova, 2022).

The formulation of climate diplomacy by each actor addresses multidisciplinary and multidimensional structures (Venturini et al., 2014). In this respect, new approaches and concepts were used to develop more efficient and unique policies for climate diplomacy. The notion of glocal is a kind of framework for a new phenomenon that has been partially used to understand climate change policies (Lee, 2014; Stoett, 2016). Glocal can be considered to be a balanced and inclusive concept that gives a holistic perspective on climate change policies in terms of sustainability. Still, there is a clear gap in climate diplomacy.

Türkiye has emerged as one of the leading countries for climate change policies, although it was a late signatory of the Paris Agreement in 2021. Türkiye has made massive efforts at various levels to contribute solutions for climate change through institutional and practiced approaches in both domestic politics and foreign policy. In this respect, the chapter aims to analyze Türkiye's climate diplomacy efforts through the glocal framework, which asserts how local policies are impactful globally. The first part will address debates about climate diplomacy vis-à-vis tools and actors. The second part will focus on defining glocal and glocal strategies, and the last part will analyze Türkiye's climate diplomacy in two dimensions: institutionalization and Zero Waste Policy.

5.2 Climate Diplomacy: Must to Do or Soft Power Tool

Flooding, droughts, shifting climatic zones, and extreme weather will have significant economic and social impacts on entire regions. Climate change is a global issue that demands a worldwide solution. This is where climate diplomacy comes in (TSKIPURISHVILI, 2022). International collaboration is critical to ensuring that all countries can appropriately adapt to the consequences of climate change. The worldwide reaction to climate change is multifaceted, intense, and sensitive. In this respect, the journey of climate diplomacy as a term goes back to 1992 when nations ratified the United Nations Framework Convention on Climate Change (UNFCCC). The UNFCCC created a climate change regime where multilateralism played a key role in facilitating policies in multidimensional areas (Popovski, 2024).

Although the UNFCCC was the cornerstone for establishing a climate change regime, the sluggish advancement of the global climate regime indicates the need for immediate action that goes beyond international climate talks. In addition, a remarkable portion of the social consequences has not been included under the UNFCCC for many years (Lefstad & Paavola, 2024). This is why foreign policymakers have become more aware of climate change in recent years.

Climate diplomacy refers to the ecosystem of policies and strategies aiming to establish the global climate change regime and to guarantee efficient ways of tackling climate change (Purdy, 2010; Hristova & Chankova, 2020). Discussions regarding establishing a climate regime clearly show that initial definitions and approaches to climate change mainly focus on international cooperation. In other words, climate diplomacy is the activity and process of developing a global climate change regime and ensuring its successful implementation. To address the internal and external hurdles to success, climate diplomacy must draw on best practices from modern diplomacy through innovative perspectives. In this respect, climate diplomacy is expanding and involves climate, disaster, and risk management (Mabey et al., 2013; Jayaram, 2022) . This rapid evolution has resulted in better integration of climate change into broader foreign policy and diplomacy, a rapid expansion of overlapping cooperation between state and non-state actors, and novel approaches to shaping a global conversation about the consequences and solutions to climate change.

Thanks to increasing awareness and inclusive approaches by actors in tackling climate change in recent years, there have been broader geopolitical discussions in climate diplomacy rather than a limited agenda (Wang & Liu, 2015). In the transformation of climate diplomacy, the capacity of each actor is also determinant to the climate regime and its success. In other words, how will effective national climate diplomacy profoundly contribute to the climate regime? Regarding contributions to the international climate regime, common goals defined through the UNFCCC involve cooperation and collaboration to solve climate change challenges. Tshering and Craft mention four important climate diplomacy factors: environmental sustainability as a foundation of development, carbon neutrality, moral authority, multilateralism, and coordinating diplomatic involvement (2016). Each factor, especially in climate diplomacy by nation-states, should be addressed regarding actors, strategies, and alignment with the current foreign policy agenda.

The literature about climate diplomacy in nation states primarily focuses on two main aspects: goal or instrument, which is embedded within soft power debates. This debate is also related to the categorization of states depending on their level of development and climate zone classifications derived from the Köppen-Geiger climate classification system, which divides climates into five main climate groups based on seasonal precipitation and temperature patterns (Torney & Cross, 2018; Skandalos et al., 2022). In other words, each actor has its own unique climate diplomacy with similarities to other actors with common goals.

Türkiye is a vulnerable country that is facing food security concerns, increased water stress, unforeseen disasters, and forest fires. This vulnerability is caused by a mix of climatic conditions, population exposure, and socioeconomic factors (Turhan et al, 2016; Türkiye-Country Climate and Development Report, 2022). Although Türkiye can be considered as a latecomer to the Paris Agreement, the country has made remarkable efforts in climate diplomacy in various ways. In a very short time, Türkiye launched series of policy initiatives both at the domestic and foreign policy levels, which aim to contribute to the climate regime as well as to protect Türkiye from the consequences of climate change. A fresh contribution is to rethink Türkiye's climate diplomacy with a multidimensional perspective, which shouldn't be limited to only traditional theories. In other words, climate diplomacy, which is composed of local and global aspects due to its nature, should be viewed with multidimensional and multidisciplinary perspectives that involve the glocal notion.

5.3 Glocal, Glocalization, and Strategies

The words glocal and glocalization are actually new in the literature and did not exist before the 1990s (Roudometof, 2015). Glocalization is derived from the Japanese concept of dochakuka, which initially meant the agricultural practice of adjusting farming operations to local conditions (Tullock, 1991). Along with interesting journey of the word glocalization, scholars like Robertson suggest the neologism highlights the interdependence of global and local processes, rather than their opposition. In simpler words, it is a holistic approach that balances global and local dimensions (Erçetin et al., 2017). It is a phenomenon that emerged as a result of the development of factors such as communication, trade, technology, and transportation, and the removal or reduction of borders between different regions of the world.

Glocalization is reflected in many areas such as economy, politics, culture, and society. The effects of glocalization include combining global and local characteristics, preserving cultural diversity, strengthening local economies, giving local governments more authority and responsibility, increasing local public participation, and greater cooperation at a global level (Açıkalın, 2023). However, globalization also has some negative effects. The negative effects of globalization include: global corporations suppressing local economies, cultural homogenization, insufficient participation of local people, exploitation of local resources by global corporations, impoverishment of local people, local producers being unable to cope with global competition, destruction of local cultures, and local governments serving the interests of global corporations (Olgun, 2007).

There is no doubt that there is a clear need for glocal policies, which bring both local and global together, to overcome global challenges, especially climate change. In this respect, policies for tackling climate change require both global and local answers. While global concerns require all nations to work together, there are also local differences that require local actors to notice, report, and respond in a timely and efficient manner (Saraseko, 2024). Collectively, there is a need to better understand the glocalization of adaptive governance in response to climate concerns. Policy-relevant knowledge of climate-related problems, the potential to integrate climate adaptation into several policy sectors for local and national capacity building, as well as just and legitimate adaptation to climate hazards, are vital for each actor (Järvikoski et al., 2007).

5.4 Understanding Türkiye's Climate Diplomacy with a Glocal Approach

As touched upon in the previous section, glocalization brings new insight to the climate policies of each actor, which is reflected in a remarkable amount of the literature (Gupta et al., 2007; Edeholt and Mainsah, 2014; Bernardini, 2024). However, there is a clear gap in the literature regarding the analysis of climate diplomacy within a glocal framework, as well as Türkiye's climate diplomacy. In this respect, the following section will touch upon Türkiye's climate diplomacy in two dimensions: institutionalization and good practices with a global impact in Türkiye.

5.4.1 Institutionalization: Building Capacity

Climate diplomacy is not solely the work of the Ministry of Foreign Affairs but should be inclusive in terms of policymaking and action taking. Institutionalization of climate-related issues in Türkiye has gained momentum, especially with international agreements and increased environmental awareness. The impact of global processes, legal and institutional arrangements, the green development movement, the role of municipalities, local governments, and civil society, and academic studies can explain the institutionalization developments.

In 1992, the UN Conference on Environment and Development (Rio Conference) held in Rio de Janeiro developed important principles for adopting an environmentally friendly management approach, and the UN Framework Convention on Climate Change (UNFCCC) was opened for signatures (UN, 1992). This convention, which entered into force in 1994, is considered the first significant step taken regarding climate change and is currently accepted by 198 countries. The UNFCCC forms the basis of the Kyoto and Paris Agreements (UN, UN Climate Change Conferences).

The UNFCCC encourages countries to reduce greenhouse gas emissions, cooperate in research and technology, and protect natural gas sinks. The Convention is based on the principle of "common but differentiated responsibilities and relative capabilities" according to each country's level of development and specific circumstances. According to this principle, countries that emitted more greenhouse gases after the Industrial Revolution assume more responsibility (UNFCCC, Article 4).

The Convention divides countries into three groups: Annex I Countries, developed countries that are obliged to reduce greenhouse gas emissions; Annex II Countries, countries that are also obliged to provide support to developing countries; and Non-Annex Countries, countries that are not under obligation but are encouraged to do so. Türkiye acquired a unique position in the UNFCCC negotiations. Türkiye, included in the Annex I and Annex II lists together with developed countries in 1992, was removed from the Annex II list with a regulation at the Marrakesh Conference in 2001 but remained in the Annex I list. This decision emphasized Türkiye's unique circumstances and drew attention to the country's position, which was different from that of other Annex I countries (MFA, UNFCC).

By joining the UNFCCC in 2004, Türkiye became the only country among Annex I countries whose "special conditions" were recognized. This decision was made considering Türkiye's characteristics, such as its level of development and environmental sensitivity.

Türkiye laid the foundations of institutional structures by becoming a party to international agreements about climate change. In 2004, Türkiye became a party to the United Nations Framework Convention on Climate Change (UNFCCC).

In 2009, Türkiye became a party to the Kyoto Protocol, but it was not obliged to reduce emissions as a developing country. Türkiye ratified the Paris Climate Agreement in 2021 and started to set concrete targets for emission reduction and combating climate change. (Directorate of Communications, 2021)

The National Climate Change Strategy coordinated by the Ministry of Environment, Urbanization, and Climate Change was approved by the Prime Ministry High Planning Council on May 3, 2010. To implement this strategy, the Climate Change Action Plan covering the years 2011–23 was prepared and implemented in July 2011. The Climate Change Action Plan aims to limit greenhouse gas emissions, adapt to climate change, and increase resilience by identifying actions appropriate to Türkiye's conditions. It also seeks to promote efforts to combat and adapt to climate change in Türkiye. The Climate Change Action Plan focuses on GHG emission control and adaptation measures, especially in priority sectors such as energy, construction, transport, industry, waste, agriculture, land use, and forestry. The Plan focuses on institutional structures, long-term cooperation, technology development, transfer, and financing, and 541 actions were identified (Ministry of Environment and Urbanization, 2020).

In 2011, another essential document, the Climate Change Adaptation Strategy and Action Plan, was published. This plan includes water resources management, agriculture and food security, ecosystem services and biodiversity, natural disaster risk management, and human health. Both action plans continue to be updated under the Ministry of Environment, Urbanization, and Climate Change, coordinated in line with medium and long-term targets for 2030 and 2053 (Directorate of Communications, 2021).

The establishment of the Turkish Environment Agency (TUÇA), which aims to prepare and publish in visual, audio, and written fields, cooperate with the press and broadcasting organs, and carry out activities such as campaigns, competitions, and promotions to raise public awareness and sensitivity, with Law No. 7261 adopted by the Turkish Grand National Assembly on December 24, 2020, and published in the Official Gazette No. 31350, has helped all environmental activities to go one step further (CSB, 31.12.2020). In 2021, the ratification of the Paris Agreement by the Grand National Assembly of Türkiye, an international agreement in which 197 countries agreed to take joint action to prevent the climate crisis, and the preparation of the Türkiye Green Deal Action Plan were two critical steps in terms of legal regulations (Özenç, 2021) (Türkiye Green Deal Action Plan, 2021, p.38).

In addition to this, the European Union plays a vital role in the fight against climate change and has set precise mechanisms for transformation through the European Green Deal. Türkiye, as a Customs Union partner with the EU, is rapidly undergoing a fundamental transformation process to fulfill the criteria of the Green Deal. This transformation could significantly impact Türkiye's trade with the EU, which is why harmonizing technical legislation is essential.

In this framework, many concrete steps were taken in the public and private sectors. Regarding Türkiye's foreign trade, the effects of policy changes on industrial, agricultural, energy, and transport policies have been considered holistically, and a roadmap for harmonization was developed considering the Customs Union relationship. Under the Ministry of Trade coordination, a Working Group was established on February 4, 2020, with the participation of many institutions, such as the Presidential Strategy and Budget Directorate, the Ministry of Environment, Urbanization and Climate Change, and the Ministry of Foreign Affairs. This group assessed the impacts of trade with the EU and prepared an Action Plan to guide Türkiye's transformation.

The Green Deal Action Plan, published in 2021, includes thirty-two targets and eighty-one actions under nine main headings to promote Türkiye's green transformation and harmonize with the EU. The plan aims to contribute to transforming global value chains, promote green investments, and support value-added production. It was also emphasized that the fight against climate change and other global issues should be addressed in conjunction with the post-Covid-19 recovery process. Türkiye is working to harmonize its climate change policies with international economic and trade policies. In this process, relevant ministries and organizations will implement environmentally friendly practices and continue their efforts to align Türkiye with global climate goals (Ministry of Trade, 2021).

Within the scope of effective combat against climate change, the Ministry of Environment, Urbanization, and Climate Change was restructured in 2021, and an independent Climate Change Presidency was established in this regard (Presidential Decree No. 85, dated 29 October 2021).

The Regulation created the legal basis for the Chief Negotiator for Climate Change, which was initiated in 2015. It was decided that the relevant Deputy Minister who reports to the Presidency would be the chief negotiator.

The Climate Change and Adaptation Coordination Board, which includes representatives of all public institutions, the business world, the educational community, and NGOs, was established. Environment and urbanization consultancies were established in international business centers such as New York, Brussels, Geneva, and Paris to ensure that climate and environmental diplomacy is carried out on-site and effectively (Ministry of Environment, Urbanization and climate change, 2021).

An updated national contribution declaration (NDC) was presented at COP27 hosted by Egypt, and a long-term climate change strategy document was published in 2024, with contributions from all relevant stakeholders.

Lastly, local governments also significantly contribute to combating climate change in Türkiye. The number of cities preparing climate change action plans has increased (e.g., Izmir, Istanbul, and Gaziantep), and environmental sustainability initiatives such as the "Zero Waste Project" have been encouraged. Civil society organizations, universities, and the private sector actively raise awareness, develop projects, and support climate policies. NGOs such as the TEMA Foundation and WWF Türkiye have made significant contributions in this field.

5.4.2 Good Practices from Local to Global: Zero Waste Project

"Zero Waste" can be defined as a resource and waste management approach based on the circularity of waste. It encourages sustainable consumption and production habits and supports the efficient use of resources. Zero waste advocates that waste can be prevented, reduced, reused, and recycled. The circular economy is an economic system that aims to minimize and eliminate waste and pollution by keeping all components and materials in use for as long as possible. This contrasts with the traditional linear economy based on a take-make-waste model. The circular economy is based on three principles: reduce, reuse, and recycle; with design for longevity and to close the loop (Directorate of Communications 2023, s.217).

The Zero Waste Project was launched in 2017 with an event organized at the Presidential Complex hosted by First Lady Emine Erdoğan (TCCB, 2017). Emine Erdoğan first sent a Zero Waste Letter to 1,000 opinion leaders, including the rectors of many universities, businesspeople, and media executives, to support the Zero Waste Project (Kılınç, 2023). To publicize the Zero Waste Project in all state institutions and society, important institutions and organizations of the state came together at a dinner meeting with representatives of non-governmental organizations working on environmental issues, from Minister of Environment and Urbanization to Turkish Airlines (TCCB, 01/23/2018).

Following this meeting, many institutions and organizations started working on Zero Waste within their structures. These activities pioneered and enabled the establishment of a zero-waste information system within the Ministry of Environment and Urbanization (Presidency Legislation Information System, Zero Waste Regulation). Immediately afterward, in 2018, regulations were made in the Environmental Law regarding charging for plastic bags, the application of recovery contribution share, the dissemination of zero waste systems, and the deposit-refund system (Hürriyet, 12/28/2018) (Dünya, 01/10/2019).

The Zero Waste project, which has been expanding daily, spread from institutions and organizations in Türkiye and was appreciated by the public, and began to receive awards from distinguished institutions and individuals worldwide. First, Viorel Gutu, UN Food and Agriculture Organization (FAO) Subregional Coordinator for Central Asia and Representative to Türkiye, presented Mrs. Emine Erdoğan with the FAO's "Zero Waste, Zero Hunger" themed award on the occasion of World Food Day on 17 October for her environmental awareness projects (Hürriyet, 01.11.2018). Afterward, developments continued unabated. Emine Erdoğan, who highly values the opinions of academic authorities, exchanged ideas about the Zero Waste Project with Professor Jeffrey Sachs, president of the Earth Institute at Columbia University, and received information from Professor Sachs about waste management models and successful practices in the world (TCCB, 09.26.2018), Subsequently, the first Zero Waste Summit was held with the theme "For a Strong Economy and Green Nature" (CSB, 11.01.2018) (TCCB, 11.01.2018). Here, the achievements of the Zero Waste project in the first year were expressed by the Minister of Environment and Urbanization Murat Kurum with the following words:

> *In the one year since the project was implemented, we collected and recycled 2.2 million tons of packaging waste, 58 thousand tons of waste electrical and electronic goods, 38 thousand tons of vegetables and 80 thousand tons of mineral waste oil, 184 thousand tons of end-of-life tires separately at their source, and from all these, we provided 3.5 billion resources to the economy in just one year. (AA, 01.11.2018)*

The Zero Waste Movement is not limited to state institutions and organizations; artists also started to support the project. The first example of this, the "From Waste to Art" exhibition, was opened by Emine Erdoğan on June 5th, World Environment Day (TCCB, 06.05.2018).

Emine Erdoğan launched another event in eighty-one provinces to keep all coastlines, lakes, rivers, and water resources clean and collect waste, especially in the context of marine cleanliness and protection of water assets (Ministry of Environment, Urbanization and Climate Change, p. 37–38). In this context, she attended the Zero Waste Blue Project Promotion Meeting launched across Türkiye (Kılınç, 2013, p. 165) (TCCB, 06.10.2019). The movement spread rapidly in Türkiye and continues to be implemented worldwide. In this context, Emine Erdoğan participated in panels and summits abroad in 2019. The first was the "Oceans: Precious Resources of Our Lives" symposium in Osakan (TCCB, 06.29.2019a). During this trip, she also met with the authorities of Kamikatsu Island, the world's first zero-waste island (TCCB, 06.292019b). This was followed by a panel discussion on "Problems and Solutions in Sustainable Waste and Resource Management: Why National Initiatives on Zero Waste Matter" (TCCB, 09.252019).

In 2021, Emine Erdoğan was also awarded the "Waste Wise Cities Global Champion" award by the United Nations Human Settlements Program (UN-Habitat) for her support of the Waste Wise Cities initiative and her contributions to the environment with the Zero Waste Project carried out under her auspices (TCCB, 06.02.2021). 2022 was a year of breathtaking developments in the Zero Waste journey. First, Emine Erdoğan was awarded the "Parliamentary Assembly of the Mediterranean Award (PAW Awards)" as the first among twenty-one countries at the 16th General Assembly of the Parliamentary Assembly of the Mediterranean held in Dubai on March 9–10, 2022 (AA, 03.18.2022).

Undoubtedly, the establishment of the Climate Change Presidency under the Ministry of Environment, Urbanization, and Climate Change was one of the biggest environmental achievements for our country in order to combat climate change, which causes profound changes in all areas from education to culture, from economy to agriculture, and from industry to urbanism. (NTV, 28.06.2022) Subsequently, a valuable contribution was the "Waste Collectors Circular," which was issued to register and authorize those who work as independent waste collectors with the "Zero Waste Information System" and to regulate the working system (CSB, 06.22.2022).

In addition, the book The World is Our Common Home was published, which includes the inspiring stories of twenty-eight environmental volunteers. It is named after the slogan "The World is Our Common Home," one of the most important slogans of the Zero Waste Project.

"This world is our common home. It is our common responsibility to keep our home beautiful and clean and to ensure that future generations can call it home." *(The World is Our Common Home, p. 37)*

On the same day as the book launch, Emine Erdoğan was presented with the "Climate and Development Leadership Award," given for the first time by the World Bank for her pioneering work on climate with the Zero Waste Project (AA, 01.06.2022) (Figure).

March 30 as International Day of Zero Waste

One of the most important milestones of the Zero Waste Project was a special meeting between Emine Erdoğan, who accompanied the president of the Republic of Türkiye, Recep Tayyip Erdoğan, to New York to attend the 77th General Assembly of the United Nations, and United Nations Secretary-General Antonio Guterres on September 19, 2022. During this meeting, possible international cooperation within the scope of the Zero Waste Project carried out in Türkiye was discussed. Following the meeting, the "Commitment to Zero Waste for the World, Our Common Home," regarding promoting and supporting Zero Waste implementation, was jointly signed (TCCB, 09.202022).

Commitment to Zero Waste includes the following statements.

"We, the undersigned:

Recognising the need to promote the efficient use of finite resources, reduce and eliminate waste generation through prevention, reduction, recycling and reuse, among others, in order to achieve the goals of the Paris Agreement and 2030 Agenda for Sustainable Development,

Acknowledging the strong link between sustainable waste management, resource efficiency and climate change,

We commit ourselves to implement sustainable waste management practices,

We pledge to promote the zero waste approach worldwide and share best practices for its wider recognition and application,

Moreover, we pledge to support initiatives, campaigns, programs, projects, and activities that promote zero waste in order to encourage a responsible waste production and consumption,

We commit to sharing the best practices and lessons learned from projects such as Zero Waste Project of Türkiye, encouraging the development of similar policies on waste management,

We call for UN Member States, UN System entities, and members of the civil society, private sector, media, as well as local authorities to align strategies, approaches and programmatic responses to help accelerate a global transition to circular economy and reduce the adverse impact of waste on climate change.(Zero Waste Commitment)

This commitment is an essential step in disseminating the Zero Waste movement launched in Türkiye under the auspices of Mrs. Emine Erdoğan at the individual level and its institutionalization at the global level. Mrs. Erdoğan has brought the commitment text, which she first signed with the UN Secretary-General, to the agenda in all her international contacts. The number of spouses of leaders and international representatives who became Zero Waste volunteers by signing the Commitment to Zero Waste has exceeded forty (AA, 12.31.2023). With the Commitment to Zero Waste, this environmental volunteer movement which began in Türkiye was carried to the United Nations.

Subsequently, Türkiye's draft resolution, "Promoting zero-waste initiatives to advance the 2030 Agenda for Sustainable Development," was submitted to the United Nations General Assembly to institutionalize cooperation and take concrete steps to promote and expand zero-waste practices. As a result, the resolution was adopted unanimously on December 14, 2022, co-sponsored by 105 countries. The resolution of the General Assembly declared March 30 as International Day of Zero Waste. The resolution calls for the UN chief to create a three-year advisory board of experts chosen for their "knowledge, experience, and expertise" to support local and national zero-waste initiatives. It also urges ongoing dialogue about zero-waste efforts, encouraging member states, UN organizations, and other international and regional bodies to adopt and implement such initiatives at all levels (Zero Waste, 12.14.2022). In January 2023, a promotional meeting was held for the "Water Efficiency Mobilization" launched by the Ministry of Agriculture and Forestry, where Emine Erdoğan stated that they were united in the promotion of a nationwide movement to protect water resources and transfer them to future generations (Ministry of Agriculture and Forestry, 01.31.2023).

The first International Zero Waste Day on March 30th was celebrated at the UN General Assembly Hall, hosted by Emine Erdoğan with the participation of the UN Secretary-General and relevant UN office and program directors, spouses of heads of state, and relevant foreign ministers. With her first speech in this program, Emine Erdoğan became the first Turkish first lady to deliver a speech in the UN General Assembly Hall (TCCB, 03.302023) (AA, 31.12.2023).

As another outcome of the resolution, an Advisory Board of Eminent Persons on Zero Waste was established under the auspices of the UN Secretary-General

(UNSG), which will operate within the United Nations. The UNSG proposed Mrs. Emine Erdoğan chair the Board within the framework of the 30th March celebrations (Zero Waste, 07.27.2023), and she accepted this offer. After joint consultations, a twelve-member board was formed with Mrs. Erdoğan as the Chairperson. The Board includes representatives such as the first ladies of Sierra Leone and Suriname, businessmen, representatives of international organizations, academics, and so on. (Figure 5.2).

A website (https://zerowastecommitment.com) was created to bring the Zero Waste Movement, which became global with the Commitment to Zero Waste, down to the individual level and to ensure the participation of every person who wants to and to open the text for international individual signatures. President Recep Tayyip Erdoğan became a Zero Waste Volunteer by providing the first signature on the website within the scope of the "Path to the Global Zero Waste Movement" event at the Turkish House in New York (TCCB, 18.09.2023a). This website has already become a medium where more than ten thousand zero waste volunteers from ninety-nine countries have signed the Global Commitment to Zero Waste, and the number of people participating from all over the world is increasing day by day (AA, 31.12.2023).

On the same day, Emine Erdoğan attended the "Path to the Global Zero Waste Movement" event at the Turkish House in New York City, to promote the opening of the Global Commitment to Zero Waste for individual signatures. The program was also attended by the spouses of heads of state, senior UN officials, and diplomats from other countries who were in New York for the 78th General Assembly of the United Nations (TCCB, 18.09.2023b).

The "World Cities Day" program, organized under the theme "Financing sustainable urban futures for all," was celebrated on October 31, 2023, with events at the Üsküdar Address Hotel in Istanbul. (UN, World Cities Day Focuses on Financing Sustainable Development). As an example of cultural diplomacy, after the program, Emine Erdoğan visited Topkapı Palace with the wives of the heads of state who came to Istanbul to attend the "31 March World Cities Day" program. (TCCB, 10.31.2023b)

In 2024, Emine Erdoğan also carried out many activities regarding the Zero Waste Project as chairperson of the UN Advisory Board of Eminent Persons on Zero Waste. Following the first meeting of the UN Advisory Board of Eminent Persons on Zero Waste, held in November 2023, the second meeting was held online on June 5. At the meeting, Emine Erdoğan stated that the preparation of guide publications for countries and organizations that want to implement Zero

Waste in cooperation with the UN had begun, and said, "We are launching our initiatives for a global Zero Waste standardization that will be valid all over the world and will unify all Zero Waste practices." The third meeting of the UN Zero Waste Eminent Persons Advisory Board was held on December 13.

The 7th anniversary of the Zero Waste Movement was commemorated with the slogan "7 Continents in 7 Years." Meanwhile, the Zero Waste Project was honored with the MEDITERRANEAN-PA award at a ceremony held in Braga, Portugal. In November, Emine Erdoğan attended the 29th Conference of the Parties (COP29) of the UN Framework Convention on Climate Change in Baku, Azerbaijan. Accompanying President Erdoğan at the World Leaders Climate Summit program organized within the scope of COP29, Emine Erdoğan met with young climate volunteers at the "Ecological Education."

5.5 Conclusion

Climate diplomacy is becoming increasingly widely employed in the international system in terms of actors and action. Although there are several definitions of climate diplomacy, the scope, goal, actor, methods, and target population are more diverse than ever. Nation-states also prioritize climate diplomacy as a foreign policy priority and a critical component of soft power by developing tailored climate diplomacy instruments and contributing to multilateral partnerships. Each actor's climate diplomacy targets interdisciplinary and complex systems. In this regard, novel ideas and concepts have been applied to create more effective and distinctive climate diplomacy.

Climate diplomacy by each actor involves interdisciplinary and complex systems. In this regard, novel ideas and concepts have been implemented to create more effective and distinctive climate diplomacy programs. The concept of glocal is a novel paradigm that has been utilized to better comprehend climate change policy. Glocal may be regarded as a balanced and inclusive concept that provides a comprehensive view of climate change strategies in terms of sustainability. Türkiye has become one of the top countries in terms of climate change policy.

In this chapter, Türkiye's climate diplomacy was analyzed within a glocal framework in two dimensions: institutionalization and good practice from local to global levels. First of all, Türkiye established both public institutions, such as the Environment Agency, as well as restructuring existing institutions like the Ministry of Environment and Urbanization. In addition to this, Türkiye developed legal frameworks and regulations related to climate change policies vis-à-vis the harmonization of various stakeholders. Türkiye facilitated institutionalization as one of the vital components of climate diplomacy from a glocal perspective. In other words, Türkiye established tailor-made, unique pillars for efficient climate diplomacy, both for soft power and to contribute to the climate regime.

Second, the Zero Waste Project is one of the leading Turkish projects that have global implications. The Zero Waste policy can be given as a concrete example of glocal policies that "Act Local Think Global." The term "zero waste" refers to a cyclical approach to resource and waste management. It fosters sustainable manufacturing and consumption behaviors, as well as resource efficiency, which started with Türkiye and became global phenomenon within only seven years, thanks to efforts of First Lady Emine Erdoğan. By 2022, 30th of March has been declared Zero Waste Day by the UN. This clearly shows that Türkiye managed to bring its local solutions to a global level, contributing to the climate regime as well as the soft power of Türkiye.

Consequently, from a glocal perspective, Türkiye's climate diplomacy has been growing in the last five years through institutionalization as well as introduction of local projects on a global scale. There is no doubt that glocalization maximizes Türkiye's visibility in the climate regime in terms of inclusiveness of policy implementation, and Türkiye has the potential to be a pioneer in this field in the coming years.

Bibliography

Açıkalın, Ş. N. (2023). Türkiye'nin Tanıtım Stratejileri. In O. Bilgin (Ed.), *Türkiye'de Şehir Markalaşması ve Küreyerel Stratejiler: Balıkesir Örneği* (pp. 89–112). Diplomasi Vakfı Yayınları.

Anadolu Agency. (2018, November 1). Çevre ve Şehircilik Bakanı Murat Kurum: Amacımız sıfır atık sistemini tüm Türkiye'ye yaygınlaştırmak. https://www.aa.com.tr/tr/turkiye/cevre-ve-sehircilik-bakani-murat-kurum-amacimiz-sifir-atik-sistemini-tum-turkiyeye-yayginlastirmak/1300113

Anadolu Agency. (2022, March 18). Sıfır Atık Akdenizin En Iyi Projesi Seçildi. https://www.aa.com.tr/tr/cevre/sifir-atik-akdenizin-en-iyi-projesi-secildi-/2539033

Anadolu Agency. (2022, June 1). Dünya Bankasından Emine Erdoğan'a 'İklim ve Kalkınma Liderlik Ödülü'. https://www.aa.com.tr/tr/gundem/dunya-bankasindan-emine-erdogana-iklim-ve-kalkinma-liderlik-odulu/2642543

Anadolu Agency. (2023, December 31). Emine Erdoğan 2023'te Gazze'de katliamın durması için çalışma yürüttü. https://www.aa.com.tr/tr/gundem/emine-erdogan-2023te-gazzede-katliamin-durmasi-icin-calisma-yuruttu/3096599

Anadolu Agency. (2024, December 31). Emine Erdoğan'ın 2024'te de öncelikli gündemi "Gazze'de barışın tesisi" ve "Sıfır Atık" oldu. https://www.aa.com.tr/tr/gundem/emine-erdoganin-2024te-de-oncelikli-gundemi-gazzede-barisin-tesisi-ve-sifir-atik-oldu/3438431

Bengisu Özenç. (2021). Türkiye Paris'e Ne Kadar Uzak?* Paris İklim Anlaşması'nda çok tartışılan, az bilinen doğrular. https://sefia.org/blog/turkiye-parise-ne-kadar-uzak-paris-iklim-anlasmasinda-cok-tartisilan-az-bilinen-dogrular/?gclid=EAIaIQobChMI8u3Xr8yIhAMVHxYGAB0raAVnEAAYASAAEgL9QvD_BwE

Bernardini, C. (2024). Regulatory tools for adaptation to "glocal" climate change. In *climate change and human systems: Innovative adaptation services as a soft-resilience strategy to tackle emerging risks* (pp. 31–70). Springer Nature Switzerland.

Bodansky, D. (2001). The history of the global climate change regime. *International Relations and Global Climate Change, 23*(23), 505.

Cumhurbaşkanlığı Mevzuat Bilgi Sistemi. Sıfır Atık Yönetmeliği. https://www.mevzuat.gov.tr/mevzuat?MevzuatNo=32659&MevzuatTur=7&MevzuatTertip=5

Dünya. (2019, October 1). Plastik Poşetlerin Ücretlendirilmesi ve Geri Kazanım Payı. https://www.dunya.com/kose-yazisi/plastik-posetlerin-ucretlendirilmesi-ve-geri-kazanim-payi/436780

Edeholt, H., & Mainsah, H. N. (2014, October). Glocal participatory design: Designerly approaches for trans-local climate change initiatives. In *Proceedings of the 13th participatory design conference: Short papers, industry cases, workshop descriptions, doctoral consortium papers, and keynote abstracts-volume 2* (pp. 75–78).

Erçetin, Ş. Ş., Potas, N., Açıkalin, Ş. N., Yilmaz, M., Kisa, N., & Gungor, H. (2017). Glocal leadership behaviors of managers working in private institutions. *Polish Journal Of Management Studies*, 16(2), 74-87.

Ghil, M., & Lucarini, V. (2020). The physics of climate variability and climate change. *Reviews of Modern Physics, 92*(3), 035002.

Gupta, J., Van Der Leeuw, K., & De Moel, H. (2007). Climate change: A 'glocal' problem requiring 'glocal' action. *Environmental Sciences, 4*(3), 139–148.

Habertürk. (2023, August 8). Dünya Şehirler Günü Programına Üsküdar Belediyesi Ev Sahipliği Yapacak. https://www.haberturk.com/ankara-haberleri/31876502-dunya-sehirler-gunune-uskudar-belediyesi-ev-sahipligi-yapacak

Hristova, A., & Chankova, D. (2020). Climate diplomacy--a growing foreign policy challenge. *Juridical Tribune/Tribuna Juridica*, 10(2).

Hürriyet. (2018, November 1). FAO Ödülü, Sıfır Atık, Sıfır Açlık. https://www.hurriyet.com.tr/gundem/fao-odulu-sifir-atik-sifir-aclik-41005901

Hürriyet. (2018, December 28). Ücretli Poşet Uygulaması Ne Zaman Başlayacak. https://www.hurriyet.com.tr/gundem/ucretli-poset-uygulamasi-ne-zaman-baslayacak-41066121

Järvikoski, T., Karjalainen, T. P., & Sulkala, T. (2007). Glocalization of the environmental discourse: Comparisons between northern Finland and northern Russia. *REALIDAD-The South Cone Journal on Social Psychology*, (4/5).

Jayaram, D., & Ajayan, R. (2022). Human security in times of climate change: Climate diplomacy for integrated action. In S. Eslamian & F. Eslamian (Eds.), *Disaster risk reduction for resilience: Disaster risk management strategies* (pp. 261–293). Springer International Publishing.

Kılınç, S. M. (2023). *Sıfır atık hareketi*. Turkuvaz Kitap.

Lee, T. (2014). *Global cities and climate change: The translocal relations of environmental governance*. Routledge.

Lefstad, L., & Paavola, J. (2024). The evolution of climate justice claims in global climate change negotiations under the UNFCCC. *Critical Policy Studies*, *18*(3), 363–388.

Mabey, N., Gallagher, L., & Born, C. (2013), The evolution of climate diplomacy and the international climate regime. In *Understanding climate diplomacy: Building diplomatic capacity and systems to avoid dangerous climate change* (pp. 21–34). E3G. www.jstor.org/stable/ resrep17706.6

Ministry of Agriculture and Forestry of the Republic of Türkiye. (2023, January 31). Su Verimliliği Seferberliği Başlatıldı. https://www.tarimorman.gov.tr/Sayfalar/Detay.aspx?TermStoreId=368e785b-af33-487d-a98d-c11d5495130b&TermSetId=2d27772a-6e08-4899-8453-5694a9eb6e6e&TermId=fb841593-eb13-4a7b-90e6-762d1dc46795&UrlSuffix=5710/Su-Verimliligi-Seferberligi-Baslatildi

Ministry of Environment, Urbanization and Climate Change of the Republic of Türkiye. (2020, December 12). Çevre Ajansının Kurulması İle Bazı Kanunlarda Değişiklik Yapılmasına Dair Kanun Resmi Gazetede yayımlandı. https://hukuk.csb.gov.tr/cevre-ajansinin-kurulmasi-ile-bazi-kanunlarda-degisiklik-yapilmasina-dair-kanun-resmi-gazetede-yayimlandi-haber-257676

Ministry of Environment, Urbanization and Climate Change of the Republic of Türkiye. (2018, November 1). Sıfır Atık Zirvesi Cumhurbaşkanlığında Gerçekleştirildi. https://www.csb.gov.tr/sifir-atik-zirvesi-cumhurbaskanligi-kulliyesi-nde-gerceklestirildi-bakanlik-faaliyetleri-25343

Ministry of Environment, Urbanization and Climate Change of the Republic of Türkiye. (2021). AB müşavirlikleri. https://ab.csb.gov.tr/musavirlikler-i-104305

Ministry of Environment, Urbanization of the Republic of Türkiye. (2022, June 22). Republic of Türkiye Climate Change Action Plan. https://cygm.csb.gov.tr/atik-toplayicilari-konulu-genelge-duyuru-426804

Ministry Foreign Affairs of the Republic Türkiye, United Nations Framework Convention on Climate Change (UNFCCC) and the Kyoto Protocol

Ministry of Trade of the Republic of Türkiye. (2021). Türkiye Green Deal Action Plan. chrome-extension://efaidnbmnnnibpcajpcglclefindmkaj/https://ticaret.gov.tr/data/60f1200013b876eb28421b23/MUTABAKAT%20YE%C5%9E%C4%B0L.pdf

Ministry of Trade of the Republic of Türkiye. (2021). Türkiye Green Deal Action Plan. https://webdosya.csb.gov.tr/db/dongusel/icerikler/turk-ye-yes-l-mutabakat-eylem-plan--20220615123955.pdf

NTV. (2022, June 28). İklim Değişikliği Başkanlığı Kuruldu. https://www.ntv.com.tr/turkiye/iklim-degisikligi-baskanligi kuruldu,xxJCmjJq8kaPScsuMvHpug

Olgun, H. (2007). Türkiye'de Büyükşehir Belediyelerine Küre-Yerelleşme (Glocalisation) ve Yerellik (Subsidiarity) Kavramları Çerçevesinde Bir Bakış. *Sosyoekonomi*, *5*(5).

Popovski, V. (2024). Climate care as a new purpose of the United Nations. In *The social consequences of climate change* (pp. 101–111). Emerald Publishing Limited.

Presidential Decree No. 85, dated 29 October 2021, https://www.resmigazete.gov.tr/eskiler/2021/10/20211029.pdf

Presidency of the Republic of Türkiye. (2017, September 26). Emine Erdoğan'ın Öncüğünde Sıfır Atık Projesi Başlatıldı. https://www.tccb.gov.tr/haberler/410/83658/emine-erdoganin-onculugunde-sifir-atik-projesi-baslatildi

Presidency of the Republic of Türkiye. (2018, January 23). Emine Erdogan Bireylerin Hayatlarinda Yapacakları Basit Degisiklikler Bizi Cok Onemli Sonuclara Goturur, https://www.tccb.gov.tr/haberler/410/89193/emine-erdogan-bireylerin-hayatlarinda-yapacaklari-basit-degisiklikler-bizi-cok-onemli-sonuclara-goturur

Presidency of the Republic of Türkiye. (2018, September 26). Emine Erdoğan Columbia Üniversitesi Dünya Enstitüsü Başkanı Prof. Dr. Sachs ile Görüştü. https://www.tccb.gov.tr/haberler/410/98823/emine-erdogan-co.lumbia-universitesi-dunya-enstitusu-baskani-prof-sachs-ile-gorustu.

Presidency of the Republic of Türkiye. (2018, November 1). Emine Erdoğan Gelecek Nesillere Kaynakları Tükenmiş bir Dünya Bırakamayız. https://www.tccb.gov.tr/haberler/410/99501/emine-erdogan-gelecek-nesillere-kaynaklari-.tukenmis-bir-dunya-birakamayiz-

Presidency of the Republic of Türkiye. (2018, June 5). Emine ErdoğanSıfır Atık Projesi Kapsamında Atıktan Sanata Sergisinin Açılışını Yaptı. https://www.tccb.gov.tr/haberler/410/94292/emine-erdogan-sifir-atik-projesi-kapsaminda-atiktan-sanata-sergisinin-acilisini-yapti

Presidency of the Republic of Türkiye. (2019a, June 29). Emine ErdoğanOsaka'da Okyanuslar Hayatımızın Değerli Kaynakları Sempozyumuna Katıldı. https://www.tccb.gov.tr/haberler/410/106844/emine-erdogan-osaka-da-okyanuslar-hayatimizin-degerli-kaynaklari-sempozyumuna-katildi

Presidency of the Republic of Türkiye. (2019, October 10). Emine Erdoğan: "Ülkemizin havasını, suyunu, toprağını temiz tutmak, istisnasız herkesin vatani görevidir".

https://www.tccb.gov.tr/haberler/410/105627/emine-erdogan-ulkemizin-havasini-suyunu-topragini-temiz-tutmak-istisnasiz-herkesin-vatani-gorevidir-
Presidency of the Republic of Türkiye. (2019, September 25). Emine Erdoğan, Kamikatsu Belediye Başkanı Hanamoto ve Sıfır Atık Akademisi yetkilileriyle görüştü. https://www.tccb.gov.tr/haberler/410/109842/emine-erdogan-new-york-ta-sifir-atik-la-ilgili-ulusal-girisimler-neden-onemli-etkinligine-katildi
Presidency of the Republic of Türkiye. (2019b, June 29). https://www.tccb.gov.tr/haberler/410/106867/emine-erdogan-kamikatsu-belediye-baskani-hanamoto-ve-sifir-atik-akademisi-yetkilileriyle-gorustu
Presidency of the Republic of Türkiye. (2021, June 2). Emine Erdoğan, çevreye olan katkılarından dolayı UN-Habitat tarafından "Atık Alanında Akıllı Şehirler Küresel Şampiyonu" ödülüne layık görüldü. https://www.tccb.gov.tr/haberler/410/128114/emine-erdogan-cevreye-olan-katkilarindan-dolayi-un-habitat-tarafindan-atik-alaninda-akilli-sehirler-kuresel-sampiyonu-odulune-layik-goruldu
Presidency of the Republic of Türkiye. (2022, September 20). Cumhurbaşkanı Erdoğan, New York'ta Küresel Sıfır Atık İyi Niyet Beyanı'na ilk imzayı attı. https://www.tccb.gov.tr/haberler/410/149483/cumhurbaskani-erdogan-new-york-ta-kuresel-sifir-atik-iyi-niyet-beyani-na-ilk-imzayi-atti
Presidency of the Republic of Türkiye. (2022, September 20). Emine Erdoğan ve BM Genel Sekreteri Guterres "Sıfır Atık Projesi" için iyi niyet beyanı imzaladı. https://www.tccb.gov.tr/haberler/410/139727/emine-erdogan-ve-bm-genel-sekreteri-guterres-sifir-atik-projesi-icin-iyi-niyet-beyani-imzaladi
Presidency of the Republic of Türkiye. (2023, March 30). Emine Erdoğan: "Çevre sorunlarında çözümün bir parçası olma kararlılığımızı sürdürüyoruz". https://www.tccb.gov.tr/haberler/410/145495/emine-erdogan-cevre-sorunlarinda-cozumun-bir-parcasi-olma-kararligimizi-surduruyoruz-
Presidency of the Republic of Türkiye. (2023, July 27). Emine Erdoğan, BM Sıfır Atık Danışma Kurulu'nun ilk toplantısına başkanlık etti. https://www.tccb.gov.tr/haberler/410/148932/emine-erdogan-bm-sifir-atik-danisma-kurulunun-ilk-toplantisina-baskanlik-etti
Presidency of the Republic of Türkiye. (2023a, October 31). Emine Erdoğan Dünya Şehirler Günü Programına Katıldı. https://www.tccb.gov.tr/haberler/410/149962/emine-erdogan-dunya-sehirler-gunu-programina-katildi
Presidency of the Republic of Türkiye. (2023b, September 18). Emine Erdoğan, New York'ta "Küresel Sıfır Atık Hareketine Doğru" programına katıldı. https://www.tccb.gov.tr/haberler/410/149504/emine-erdogan-new-york-ta-kuresel-sifir-atik-hareketine-dogru-programina-katildi
Presidency of the Republic of Türkiye. (2023c, October 31). Emine Erdoğan Devlet Başkanları Eşleri ile Topkapı Sarayını Gezdi. https://www.tccb.gov.tr/haberler/410/149969/emine-erdogan-devlet-baskanlarinin-esleriyle-topkapi-sarayi-ni-gezdi

Purdy, J. (2009). The politics of nature: Climate change, environmental law, and democracy. *Yale LJ*, *119*, 1122.

Purdy, J. (2010). The politics of nature: Climate change, environmental law, and democracy. *The Yale Law Journal*, *119*(6), 1122–1209. http://www.jstor.org/stable/20698322

Robertson, R. (1995). Glocalization: Time–space and homogeneity–heterogeneity. In M. Featherstone, S. Lash, & R. Robertson (Eds.), *Global modernities* (pp. 25–54). Sage.

Roudometof, V. (2015). The glocal and global studies. *Globalizations*, *12*(5), 774–787. https://doi.org/10.1080/14747731.2015.1016293

Saraseko, T. (2024) Upscaling climate change—glocalizing governance. In *Constructing global challenges in world politics* (pp. 149–174). Routledge.

Sending, O. J., Øverland, I., & Hornburg, T. B. (2019). Climate change and international relations. *Journal of International Affairs*, *73*(1), 183–194.

Skandalos, N., Wang, M., Kapsalis, V., D'Agostino, D., Parker, D., Bhuvad, S. S., & Karamanis, D. (2022). Building PV integration according to regional climate conditions: BIPV regional adaptability extending Köppen-Geiger climate classification against urban and climate-related temperature increases. *Renewable and Sustainable Energy Reviews*, *169*, 112950.

Sprinz, D., & Luterbacher, U. (1996). *International relations and global climate change.* MIT Press.

Stoett, P. J. (2016). Human rights, glocal ecopolitics, and contested landscapes of sovereignty. In *Negotiating sovereignty and human rights* (pp. 181–197). Routledge.

Thuiller, W. (2007). Climate change and the ecologist. *Nature*, *448*(7153), 550–552.

Torney, D., & Cross, M. A. K. D. (2018). Environmental and climate diplomacy: Building coalitions through persuasion. In C. Adelle, K. Biedenkopf, & D. Torney (Eds.), *European Union external environmental policy: Rules, regulation and governance beyond borders* (pp. 39–58). Palgrave Macmillan.

TRT. (2022, June 1). 'Dünya Ortak Evimiz' kitabı bugün okuyucuyla buluşuyor. https://www.trthaber.com/haber/gundem/dunya-ortak-evimiz-kitabi-bugun-okuyucuyla-bulusuyor-684646.html

Tshering, D., & Craft, B. (2016). Engaging effectively in climate diplomacy: Policy pointers from Bhutan. *International Institute for Environment and Development.* available at: www.jstor.org/ stable/resrep02617

Tskipurishvili, A. (2022). *New climate diplomacy under the paris climate agreement.* Geneva.

Tullock, S. (Compiler). (1991). *Oxford dictionary of new words.* Oxford University Press.

Turhan, E., Cerit Mazlum, S., Şahin, Ü., Şorman, A. H., & Cem Gündoğan, A. (2016). Beyond special circumstances: climate change policy in Turkey 1992–2015. *Wiley Interdisciplinary Reviews: Climate Change*, *7*(3), 448–460.

Türkiye's Zero Waste Journey for a Habitable World, Publications by the Presidency's Directorate of Communications, 2023.

Türkiye'nin Yeşil Kalkınma Devrimi, Publications by the Presidency's Directorate of Communications, 2021.

Tykhomyrova, Y. B. (2022). The concept of climate diplomacy in modern scientific discourse. ScientificWorld-NetAkhatAVKarlsruhe.

UNFCCC, United Nations Framework Convention on Climate Change, Article 4. https://unfccc.int/resource/ccsites/zimbab/conven/text/art04.htm

United Nations. Cities Day. https://www.un.org/en/observances/cities-day

United Nations. World Cities Day Focuses Financing Sustainable Development. https://www.un.org/en/delegate/world-cities-day-focuses-financing-sustainable-development

United Nations, United Nations Conference on Environment and Development, Rio de Janeiro, Brazil, 3–14 June 1992, https://www.un.org/en/conferences/environment/rio1992

United Nations. UN climate change conferences. https://www.un.org/en/climatechange/un-climate-conferences,

Wang, W., & Liu, Y. (2015). Geopolitics of global climate change and energy security. *Chinese Journal of Population Resources and Environment*, 13(2), 119–126.

Venturini, T., Baya Laffite, N., Cointet, J. P., Gray, I., Zabban, V., & De Pryck, K. (2014). Three maps and three misunderstandings: A digital mapping of climate diplomacy. *Big Data & Society*, *1*(2), 2053951714543804.

World Bank Group. 2022. *Türkiye country climate and development report*. CCDR Series. World Bank. http://hdl.handle.net/10986/37521

Worwood, V. A. (2016). *The complete book of essential oils and aromatherapy, revised and expanded: over 800 natural, nontoxic, and fragrant recipes to create health, beauty, and safe home and work environments*. New World Library.

Zero Waste Commitment. https://zerowastecommitment.com

Zero Waste. (2022, December 14). UN adopts zero waste resolution presented by Türkiye.http://www.zerowaste.gov.tr/en/library/news/un-adopts-zero-waste-resolution-presented-by-turkiye

Zero Waste. (2023, July 27). Emine Erdoğan, BM Sıfır Atık Danışma Kurulu'nun ilk toplantısına başkanlık etti. https://sifiratik.gov.tr/kutuphane/haberler/emine-erdogan-bm-sifir-atik-danisma-kurulu-nun-ilk-toplantisina-baskanlik-etti

Zero Waste. (2021). Emine Erdoğan, BM'nin Küresel Amaçlar Eylem Ödülü'ne layık görüldü. https://sifiratik.gov.tr/kutuphane/haberler/emine-erdogan-bm-nin-kuresel-amaclar-eylem-odulu-ne-layik-goruldu

6

A Challenging Factor of Climate Diplomacy

Climate Migration

Mithat Ekiciand Sinem Sinangil

6.1 Introduction

Climate change has become an undeniable global crisis, profoundly reshaping environmental, social, and economic systems. Among its far-reaching consequences, climate-induced migration has emerged as one of the most pressing challenges, posing complex dilemmas for international governance and diplomacy. The rise in greenhouse gas emissions has triggered global warming, exacerbating critical issues such as water scarcity, declining agricultural productivity, extreme weather events, and drought. These environmental crises have severely impacted living conditions, particularly in developing countries, leading to the displacement of millions and sparking significant migration flows. Such migration movements have posed complex diplomatic challenges for nations.

This study critically examines the impact of climate change-induced migration on international diplomacy, emphasizing the diplomatic challenges and policy responses at both bilateral and multilateral levels. By assessing Türkiye's strategic role as both a transit and destination country, the study provides an in-depth understanding of how migration diplomacy is shaped by environmental factors. Climate migration is not only driven by environmental factors but is also interconnected with security concerns, economic inequalities, and political conflicts, giving rise to multifaceted diplomatic issues. Given its strategic geographical location, Türkiye plays a dual role as both a transit and destination country, making it a focal point for addressing climate migration challenges. This situation presents a significant test for Türkiye's climate diplomacy.

This study employs a mixed-methods research approach to examine Türkiye's migration policies and climate diplomacy strategies. The quantitative analysis is based on datasets from international organizations such as the United Nations High Commissioner for Refugees (UNHCR), the International Organization for Migration (IOM), and the World Bank. These data are used to evaluate the scale of climate-induced migration affecting Türkiye and its broader implications for international relations. Alongside this, a qualitative content analysis has been conducted to assess Türkiye's diplomatic engagements and international agreements concerning migration and climate diplomacy. By combining empirical data with an in-depth examination of diplomatic initiatives, this study offers a nuanced perspective on Türkiye's strategic approach to managing migration and its role in international cooperation efforts.

Climate migration has emerged as a critical issue in diplomatic negotiations, with Türkiye's role becoming increasingly pivotal in this process. As both a transit and destination country, Türkiye faces immense migration pressures, positioning it as a key actor in global climate migration dynamics. This has made climate diplomacy an essential component of Türkiye's relations with international organizations such as the United Nations (UN), the European Union (EU), North Atlantic Treaty Organization (NATO), and the Organization of Turkic States (OTS). Türkiye's experiences with refugee crises provide valuable lessons for managing future climate migration on an international scale.

This study thoroughly examines Türkiye's challenges related to climate-induced migration and the implications for international diplomacy. While climate diplomacy will require the development of more comprehensive policies moving forward, the experiences of countries like Türkiye offer crucial insights into effective climate migration management. The success of climate diplomacy will ultimately depend on robust international cooperation and collective action, with Türkiye's role being central to shaping global migration governance.

By combining empirical data with an in-depth examination of diplomatic initiatives, this study offers a nuanced perspective on Türkiye's strategic approach to managing migration and its role in international cooperation efforts. The following sections will explore the historical evolution of diplomacy, the theoretical foundations of migration diplomacy, and Türkiye's position in this evolving global landscape.

6.2 The Transformation of Diplomacy: Historical Foundations and Contemporary Challenges

Diplomacy refers to the peaceful management of relations between states under circumstances where diplomatic functions are fully operational (Barratt, 1985). It can be defined as a strategic approach that nations employ in their foreign policy. Just as foreign policy is recognized as a subfield of international relations, diplomacy emerges as a core mechanism for executing these policies. According to scholars, diplomacy is a highly specialized discipline with the primary objective of conflict prevention and the maintenance of peace. It serves as a tool for efficient communication and conflict resolution in international relations, regardless of race, language, or religion.

In contemporary discourse, diplomacy is conceptualized in two distinct forms. The first, traditional diplomacy, typically encompasses formal negotiations and agreements conducted through a country's Ministry of Foreign Affairs and diplomatic missions. This represents direct, interstate engagements. The second, broad diplomacy, extends beyond state-to-state interactions and involves international organizations such as the United Nations (UN), the World Health Organization (WHO), and the European Union (EU) (Abdurrahmanlı, 2021). Headley Bull, an Australian professor of international relations, defines diplomacy as "the conduct of relations between states and other actors in world politics through official policies and peaceful methods." (Bull, 1932).

Diplomacy can be categorized into two historical phases: classical diplomacy and modern diplomacy. Classical diplomacy, practiced by ancient civilizations such as Greece, Rome, Egypt, and the Hittites, was primarily conducted through ad-hoc diplomatic engagements (Avar, 2016). In contrast, modern diplomacy emerged after the French Revolution (1789) and evolved significantly through key diplomatic milestones such as the Congress of Vienna (1815) and the Hague Peace Conferences (1899, 1907) (Abdurrahmanlı, 2021). Globalization and technological advancements have fundamentally transformed the practice of diplomacy, shifting it from a state-centric model to a more decentralized and dynamic framework. Today, diplomacy is no longer confined to official state representatives but extends to a broader network of actors, including multinational corporations, civil society organizations, and even influential individuals. The rise of digital diplomacy, facilitated by social media, and real-time communication tools, has further altered diplomatic engagements, enabling states to shape international narratives and influence global public opinion instantaneously. In the modern international order, media, public opinion, civil society organizations, international institutions, and individual actors have become pivotal forces, directly influencing decision-makers (Directorate of Communications, 2022).

The modern era of diplomacy, shaped by the post-French Revolution period and the Congress of Vienna, is often referred to as preventive diplomacy. This concept was formally introduced by UN Secretary-General Dag Hammarskjöld in his 1959 annual report, where he described it as an approach aimed at safeguarding international peace and security through proactive conflict prevention measures. He linked this concept to nuclear disarmament and the strengthening of economic ties between states (Avar, 2016). The Kyoto Protocol and the Paris Climate Agreement, which are among the cornerstone diplomatic initiatives in global climate change governance, serve as notable examples of this diplomatic evolution.

With globalization, diplomacy has expanded into multiple interconnected spheres, reflecting the increasing complexity of international relations. Modern diplomacy now encompasses various forms, including multilateral and bilateral diplomacy, public diplomacy, environmental diplomacy, humanitarian diplomacy, and coercive diplomacy, among others. These evolving diplomatic practices highlight the necessity for states and non-state actors to adopt flexible and multi-layered approaches in addressing global challenges (Avar, 2016).

In today's interconnected world, diplomacy has evolved beyond its traditional function of managing interstate relations to become a key instrument of global governance. Complex challenges such as climate change, global health crises, cybersecurity, and migration necessitate diplomatic approaches that go beyond conventional, state-centered interactions (Alver, 2024). In response, diplomacy now involves a broader range of actors, including international organizations, multinational corporations, civil society groups, and even individuals with significant influence in global affairs.

The rapid expansion of digital diplomacy has fundamentally altered diplomatic engagements, shifting them from traditional closed-door negotiations to a more public and data-driven approach. Governments now utilize artificial intelligence, big data analytics, and cyber diplomacy tools to enhance their foreign policy strategies. Social media platforms, in particular, have become powerful instruments for real-time diplomatic communication, crisis management, and international reputation building. For instance, platforms like Twitter and Facebook have been instrumental in shaping global narratives during diplomatic conflicts and international crises. Social media platforms, in particular, have reshaped diplomatic communication, allowing states to interact with global audiences in real time and influence public discourse in ways that were previously unimaginable (Alver, 2024).

One of the defining features of modern diplomacy is its multi-layered and dynamic nature, extending beyond traditional state-driven negotiations. The growing influence of multi-track diplomacy highlights the increasing involvement of non-state actors, including local governments, academic institutions, multinational corporations, and civil society organizations. This shift has led to a more flexible and responsive diplomatic landscape, where global challenges such as climate change, migration, and security threats require broader cooperation. As a result, diplomacy has become more inclusive, integrating various stakeholders to develop sustainable and effective solutions in an increasingly interconnected world.

6.3 Rethinking Diplomacy: Public and Migration Diplomacy in Theoretical Perspective

Although public diplomacy has been practiced under different names throughout history, its conceptual foundation was established in 1965 by Edmund Gullion, the Dean of the Fletcher School of Law and Diplomacy at Tufts University. According to Gullion, public diplomacy encompasses areas of international relations that go beyond traditional diplomacy. It involves fostering positive relations between states and non-state actors with foreign publics (Directorate of Communications, 2022).

Public diplomacy is a crucial instrument through which governments engage with foreign and general publics to shape perceptions and garner support for their foreign policy objectives. Unlike traditional diplomacy, which primarily involves direct state-to-state negotiations, public diplomacy operates through civil society organizations, international institutions, and influential individuals. Its primary goal is to influence public opinion and international stakeholders, thereby enhancing a country's global image and securing strategic advantages in diplomatic engagements (Çiçek & Turan, 2023).

Migration diplomacy has become a key instrument in international relations, enabling states to manage and influence cross-border population movements. Unlike traditional migration policies, which focus on national border control and labor market regulations, migration diplomacy is deeply intertwined with broader geopolitical, economic, and security concerns. Through diplomatic negotiations, bilateral agreements, and multilateral frameworks, states leverage migration diplomacy to strengthen alliances, negotiate economic benefits, and address security challenges (Koç, 2024).

Migration diplomacy involves a variety of diplomatic measures that states use to manage cross-border population movements. These include bilateral and multilateral agreements, migration policies tailored to specific nationalities, guest worker programs, deportation policies, and repatriation agreements. Through these mechanisms, states seek to balance economic needs, security concerns, and humanitarian responsibilities, making migration diplomacy a critical aspect of international relations. Forced migration, labor migration, and diasporas are key components of migration diplomacy. For instance, following the Arab Spring, the migration of Syrian refugees to Türkiye and Jordan significantly influenced diplomatic relations between these countries and the European Union, as the latter sought diplomatic mechanisms to curb the influx of refugees into Europe (Argalı, 2021).

Theoretical discussions on migration diplomacy suggest that it is not merely a reactive policy tool but rather a proactive mechanism through which states exert influence. Scholars categorize migration diplomacy into three main approaches: migration as leverage, where states use migration pressures to extract concessions from other states; migration as a soft power tool, where a state's treatment of migrants influences its international reputation; and migration as a foreign policy instrument, where migration-related agreements serve as a means of strengthening alliances or managing security concerns (Koç, 2024; Cull and Rodríguez Jiménez, 2015).

Voluntary migration is generally directed toward countries with strong economies, democratic institutions, effective governance, and high-quality education and healthcare systems. The United States, Canada, Germany, Japan, the United Kingdom, and Australia rank among the top destinations for voluntary migrants (Çiçek and Turan, 2023). However, states experiencing large-scale migration inflows also employ migration diplomacy to negotiate financial or strategic support from international organizations and other states, thereby incorporating migration governance into their broader foreign policy agenda (Çiçek and Turan, 2023).

Forced migration, on the other hand, results from wars, totalitarian regimes, political oppression, natural disasters, and environmental crises (Çiçek and Turan, 2023). The securitization of migration has led to an increased use of migration diplomacy as a means to mitigate potential threats posed by large refugee movements. In many cases, states employ migration diplomacy not only to manage incoming refugee flows but also to exert influence on neighboring or donor countries through migration policies (Koç, 2024).

Governments that adopt inclusive policies toward forced migration often gain international recognition, as such actions reflect a commitment to humanitarian values. Türkiye's "open-door policy" for Syrian refugees serves as a notable example, aligning with international humanitarian law. However, beyond its humanitarian dimension, Türkiye's migration diplomacy has taken on a strategic character, particularly in negotiations with the European Union. The EU-Türkiye Migration Deal demonstrates how migration policies can function as a diplomatic tool, enabling states to negotiate political and economic concessions within broader international agreements (Argalı, 2021). This demonstrates how migration policies can be strategically utilized as a bargaining tool in international negotiations, allowing states to secure political, economic, or security-related concessions from other actors.

6.4 Climate Diplomacy and Migration: Turkiye's Strategic Role and Global Challenges

6.4.1 Historical Context and Global Environmental Challenges

Throughout history, humanity's interaction with nature has shaped civilizations, economies, and societies. However, the Industrial Revolution marked a turning point, as rapid mechanization and technological advancements led to severe environmental consequences. Unchecked industrial growth, population expansion, and resource depletion have intensified global challenges such as climate change, biodiversity loss, and water scarcity. These crises are no longer confined to ecological concerns but have profound economic, political, and social implications. One of the most pressing consequences of climate change is forced migration, as millions are displaced due to extreme weather events and environmental degradation. In this evolving global landscape, climate diplomacy has emerged as a crucial instrument for addressing the growing challenges of climate-induced migration and fostering international cooperation.

With the Industrial Revolution, the world entered an era of mechanization and technological advancement, which led to severe environmental pollution. Population growth and ecological degradation have resulted in global challenges such as water scarcity, biodiversity loss, nuclear testing, and climate change, making environmental issues a critical part of the international agenda. These crises have negatively affected human life, increasing global carbon emissions, worsening air pollution, and disrupting food supply chains (Arar, 2024). The intensification of droughts, floods, and desertification, particularly in regions like the Saharan expansion and the drying of major river systems, has exacerbated these challenges, particularly in areas such as the Mediterranean Basin, which is becoming more vulnerable to arid conditions (Birleşmiş Milletler, 2015).

6.4.2 The Rise of Climate Diplomacy: From the Industrial Revolution to Global Action

The development of environmental policies gained momentum in the 1960s, as awareness of global environmental issues increased. The thinning of the ozone layer was one of the first major issues that captured global attention, prompting international action. This period marked the beginning of diplomatic environmental initiatives, culminating in the 1972 Stockholm Conference on the Human Environment, the first major international effort to address environmental concerns. The conference played a crucial role in shaping modern environmental policies, leading to a significant increase in academic studies and international initiatives in the field (Birleşmiş Milletler, 2015).

Over the following decades, environmental diplomacy evolved into a critical field of international relations. The 1992 Rio Conference, known as the Earth Summit, played a pivotal role in introducing climate change as a major global issue, laying the groundwork for future climate governance frameworks. Following this, the Kyoto Protocol, adopted in 1997 and entering into force in 2005, became the first legally binding international treaty aimed at reducing greenhouse gas emissions. Later, the Paris Climate Agreement, adopted in 2015, sought to limit global warming to well below 2 degrees Celsius, with an aspirational target of 1.5 degrees Celsius (ÇŞB, 2018). The United Nations Sustainable Development Goals (SDGs), particularly Goal 13: Climate Action, emphasize the urgency of international cooperation in combating climate change and protecting vulnerable populations (Birleşmiş Milletler, 2015). However, despite these diplomatic advancements, climate governance mechanisms remain insufficient, particularly in addressing the complex relationship between climate change and forced migration (Sipahioğlu, 2022).

By the 1980s, the term "environment" was no longer sufficient to encompass the complexity of the issue. Environmental problems were increasingly linked to poverty, economic development, and resource management, leading to the emergence of the sustainable development paradigm. In 1987, the Brundtland Report ("Our Common Future"), published by the World Commission on Environment and Development, called for integrating sustainability into global policies, defining sustainable development as "meeting the needs of the present without compromising the ability of future generations to meet their own needs." (Arar)

6.4.3 *The Role of Climate Diplomacy in Addressing Climate-Induced Migration*

As climate change accelerates, extreme weather events—such as droughts, hurricanes, and rising sea levels—are increasing migration pressures worldwide. Studies indicate that the world's poorest 50 percent of the population contributes only 8 percent of total carbon emissions but bears the brunt of climate-related disasters. In contrast, the wealthiest 10 percent of the global population is responsible for nearly 50 percent of carbon emissions while being the least affected by the crisis (İstanbul Planlama Ajansı, 2024).

Therefore, it is expected that climate change will have a more severe impact on poorer countries, with its anticipated effects being more pronounced. This is because contributing to resilient urbanization and adaptation efforts to climate change requires substantial financial costs. Conditions such as droughts, wildfires, floods, biodiversity loss, and rising sea levels will affect underdeveloped countries more harshly. The livelihoods in these countries are primarily derived from natural environments, which are the most affected by the conditions caused by climate change. The deterioration of these natural and familiar environmental conditions has already significantly disrupted livelihoods and triggered internal and regional migrations. However, the degradation of natural environments, combined with the inadequacy of mitigation and adaptation measures, seems likely to initiate intercontinental migration as well.

Developing nations, especially those with limited financial and infrastructural capacity, are disproportionately impacted by climate change. For instance, countries like Bangladesh, where 15 million people could be displaced due to rising sea levels in the Ganga-Brahmaputra Delta, are expected to face severe migration challenges in the coming years (An, Turp, and Kurnaz, 2021). Additionally, areas such as the Mekong Delta, Irrawaddy River Basin, and the Nile Delta are identified as hotspots for forced migration due to sea-level rise, desertification, and water scarcity (IPCC, 2023). Without adequate climate adaptation and resilience strategies, these regions will face increasing levels of internal, regional, and transcontinental migration. According to World Bank projections, by 2050, approximately 216 million people could be forced to migrate due to climate change-related factors, including rising sea levels, desertification, and extreme weather conditions (An, Turp, and Kurnaz, 2021).

According to IPCC reports, some areas may become uninhabitable due to water shortages, agricultural collapse, and extreme heat (IOM, 2024). Countries such as Tuvalu, Kiribati, and the Maldives, which are highly vulnerable to sea-level rise, are already considering land purchases for relocation. In the Nile Delta, a rise of just 1 meter could result in the displacement of millions of people, especially those living in cities like Alexandria (An, Turp, and Kurnaz, 2021). However, the governance of climate migration remains largely underexplored, with few international mechanisms in place to manage the movement of displaced populations. Unlike political refugees, climate migrants currently do not fall under the protection of the 1951 Refugee Convention, leaving them in a legal limbo. The absence of an internationally recognized status for climate-induced displacement further complicates the response mechanisms of states and international organizations (IPCC, 2023).

Climate change is creating global impacts across a wide range of areas, from physical and mental health issues to food and water insecurity. Extreme temperatures and weather events result in fatalities, trauma, and respiratory problems, while certain diseases are becoming more widespread and spreading to new regions. People in vulnerable areas are fifteen times more at risk from these impacts. Additionally, social inequalities and marginalization further exacerbate this vulnerability. Sectors such as agriculture, fishing, and aquaculture, which rely heavily on natural conditions, have been severely affected by climate change impacts like drought, temperature increases, and ocean acidification. As a result, food production and access have been disrupted, and millions of people face acute food insecurity and malnutrition (İstanbul Planlama Ajansı, 2024).

The IPCC and other research centers predict that some regions will become uninhabitable due to climate change, forcing people to migrate. However, there is insufficient research on how to manage these migrations, where people will go, and how to address the multifaceted humanitarian issues arising. It is anticipated that the climate crisis will disproportionately affect areas with weak infrastructure. Therefore, it is crucial for governments to take the existing research gaps into account in future climate change policies. Moreover, more scientific and comparative research should be encouraged on the potential impacts of environmental changes on migration crises. Local governments can reduce migration risks by increasing green spaces and improving social services to combat flood disasters.

Climate change, along with rising temperatures, poses serious threats to food security, ecosystems, and human life. The increase in temperatures will have a greater impact on poorer countries, leading to global economic losses, limiting state finances, and slowing economic growth. This will increase internal, regional, and ultimately transcontinental migration.

In this process, especially women, children, the elderly, people with disabilities, and the poor are more vulnerable to the devastating effects of the climate crisis. The poor quality of the areas they live in, inadequate access to healthcare, and economic deprivation make these groups more vulnerable to crises. For instance, when examining migrating communities, it has been found that approximately 80 percent of these groups are women and children, and in addition to the challenges women face in accessing healthcare, they are more likely to encounter risks of sexual and physical violence (UN Women, 2022).

Climate change adaptation activities could be effective in reducing risks to both nature and human populations. Particularly in poor and migration-prone countries and regions, climate diplomacy prioritizing climate finance and related climate justice should focus on water management, farm diversification, restoring natural forests and wetlands, recognizing indigenous peoples' rights, and restoring ecosystems to protect biodiversity and human populations. Additionally, planning approaches that prioritize equity and justice, while considering climate risks, could prevent potential migrations on-site and enable preventive diplomacy that avoids creating new global conflict zones. Furthermore, providing support to displaced individuals and strengthening the capacities of host communities could offer broader options for safe and regular mobility (UNHCR, 2024).

Climate-induced migrations, which remain ambiguous in international law, refer to displacements caused by the climate crisis. For example, the Mitch and Katrina hurricanes led to the migration of hundreds of thousands of people in the United States. Countries severely affected by climate change, including India, Pakistan, China, and many Asia-Pacific countries, are expected to experience an increase in climate-induced migrations. In Africa, agricultural capacities are anticipated to decrease significantly. For instance, agricultural production in the equatorial belt is projected to decrease by up to 60 percent by 2080. As a result, food crises are expected to become widespread, and hundreds of millions of people will face malnutrition issues. By 2050, the number of migrants is projected to reach 263 million. However, determining how many of these migrants will be climate-induced will be difficult. Moreover, large cities (e.g., Miami, New York, Tokyo) are metropolitan areas that could be severely impacted by climate-related extreme weather events (Kentli, 2022).

The climate crisis is increasing displacement and making life even more difficult for those already forced to flee. Climate change and displacement are increasingly interconnected. As extreme weather events and environmental conditions worsen with global warming, they contribute to multiple overlapping crises that threaten human rights, increase poverty and the loss of livelihoods, strain peaceful relations between communities, and ultimately create conditions for more forced displacement. Today, the majority of people displaced by persecution, conflict, and violence live in countries that are highly vulnerable and unprepared to adapt to climate change. The majority of displaced populations live in overcrowded camps or informal settlements with limited access to basic services or infrastructure, making them highly vulnerable to climate hazards such as floods, droughts, storms, and heatwaves while living in foreign geographies. Moreover, the climate crisis is disrupting livelihoods and making it harder for displaced people to become self-sufficient. Climate effects also escalate tensions and conflicts over vital resources, such as water, fuel, and arable land, and threaten the potential for peaceful coexistence between displaced populations and host communities. Addressing climate change as the primary cause of displacement would be a crucial approach to breaking this cycle and finding lasting solutions.

Demographic change is increasingly shaping our future. According to current trends, the working-age population in developing countries is expected to increase by 552 million by 2030, and these countries will need to create enough jobs to achieve poverty reduction and growth targets. These projections show a significant increase in migration pressures over the next decade, especially from South Asia and Sub-Saharan Africa (UNCHR, 2024).

Migration pressures may increase, particularly in low-income and climate-vulnerable nations, as the economic costs of adaptation become increasingly challenging. Countries with limited resilience to climate change, like those in Sub-Saharan Africa and South Asia, are expected to experience escalating migration flows as a result of environmental degradation (An, Turp, and Kurnaz, 2021). Many states lack climate-resilient infrastructure, early warning systems, and effective adaptation policies, which further exacerbate migration challenges. The increasing frequency of extreme climate events, such as those experienced in regions like the Horn of Africa and the Caribbean, will likely lead to surges in climate refugees, putting additional pressure on neighboring countries (IOM, 2024). As a result, climate diplomacy is becoming an increasingly important tool in managing forced displacement, ensuring that affected populations receive international support and protection.

Migration pressures are expected to grow, especially in low-income and climate-vulnerable countries, where the costs of adapting to climate change are rising. Regions like Sub-Saharan Africa and South Asia, with limited resources, will likely see increasing migration as a result of environmental degradation. The lack of infrastructure, early warning systems, and effective climate adaptation strategies in many countries will only make this situation worse. Furthermore, extreme climate events like droughts, floods, and storms are predicted to become more frequent, intensifying migration flows and placing additional pressure on neighboring countries. Climate diplomacy is becoming increasingly important in managing this forced displacement and ensuring affected populations receive the support they need.

To address this crisis, global collaboration is essential. Climate-induced migration is a complex issue that requires international coordination. Currently, there is no clear legal framework to protect climate migrants, leaving them vulnerable. There needs to be a unified effort at the international level to recognize the rights of climate migrants and develop policies that address the root causes of migration, such as climate adaptation programs.

Rich and developed nations, which contribute significantly to global emissions, have a responsibility to help fund solutions in vulnerable countries. This cooperation could help reduce the need for migration and provide support to displaced populations. International cooperation, with clear guidelines and shared responsibility, is crucial to ensuring that migration caused by climate change does not turn into a crisis but instead is managed with fairness and compassion.

6.4.4 Türkiye's Approach to Climate Diplomacy: Key International Agreements

The United Nations Framework Convention on Climate Change (UNFCCC) was adopted in 1992. In this agreement, Annex I and Annex II countries were defined. In line with this convention, at the seventh Conference of the Parties

(COP) meeting, a decision was made that Türkiye would remain an Annex I country, while being removed from Annex II. Additionally, Türkiye's position as an Annex I country was redefined to reflect its different status (ÇŞB, 2018). Annex I countries include forty-two nations and the EU. As of 1992, OECD member countries, including Türkiye, were categorized into the first group. The second group consists of countries from the former Eastern Bloc, which were in the process of transitioning to a market economy. Annex II includes twenty-three countries and the EU. The remaining 154 countries, classified as non-Annex countries, do not bear any obligations. By using its position in Annex I, Türkiye successfully lobbied at COP16 in Mexico in 2010 to secure access to funding, capacity-building, and technology transfer opportunities, which were previously available only to Annex I countries but had not been accessible to Türkiye (MFA, 2020).

In 1992, despite being an OECD member, Türkiye was automatically classified as an Annex II country, but in 2001, it lobbied to have this classification removed, and in 2010, it successfully defined its unique position within Annex I. These interventions are seen as significant actions in climate diplomacy. Experts have assessed that these interventions have produced both positive and negative consequences (MFA, 2020).

The Kyoto Protocol, the first international climate agreement, was adopted in 1997 and entered into force in 2005 after several diplomatic negotiations. Through the protocol, the Intergovernmental Panel on Climate Change (IPCC) regularly publishes global climate assessment reports, which are part of the international political consensus and agreements (Arıkan and Özsoy, 2008). The protocol has 191 signatories, and Türkiye became a party in 2009 (MFA, 2020). However, Türkiye did not join the Annex B list of countries that have defined quantified emission reduction targets under the protocol. Therefore, Türkiye was not subjected to any emission reduction or greenhouse gas mitigation obligations during the two commitment periods from 2008–12 and 2012–20 (ÇŞB, 2018). Türkiye's approach to avoiding binding commitments in the Kyoto Protocol can be considered an example of its climate diplomacy, similar to how the victorious powers after the Second World War avoided responsibility.

As the obligations of the Kyoto Protocol ended in 2020, the Paris Climate Agreement, signed in 2016 and ratified by the Turkish Grand National Assembly in 2021, came into force. The Paris Agreement set the global temperature increase goal to remain well below 2 degrees Celsius, with an aspiration of limiting it to 1.5 degrees Celsius. The agreement came into effect on November 4, 2016, after fifty-five countries, representing 55 percent of global CO_2 emissions, ratified it. The primary goals set by the Paris Agreement include securing food security, addressing financial needs, and continuing development while protecting the future. The responsibilities defined in the Kyoto Protocol were based on the principle of "common but differentiated responsibilities and respective capabilities," where each country fights climate change according to its capacity. Parties were required to submit their nationally determined contributions (NDCs) every 5 years. Türkiye submitted its Intended Nationally Determined Contribution (INDC) to the UNFCCC Secretariat on September 30, 2015. Türkiye emphasized two important issues within the Paris Agreement: "Türkiye's equal treatment in terms of access to financial and technological support compared to countries with similar conditions" and "the necessity to acknowledge the impossibility of absolute emission reductions due to economic growth, population increase, and other indicators" (ÇŞB, 2018). Türkiye's approach to the Paris Agreement again reflects a diplomatic strategy, with examples similar to its stance in the Kyoto Protocol. Here, the principle of "common but differentiated responsibilities and respective capabilities" is aligned with Türkiye's national realities.

Specifically, it highlighted its dependence on external energy resources and its strategic location as a migration transit point, emphasizing its historical and current minimal contribution to global emissions, thereby gaining time for mitigation and adaptation efforts, which is a good example of climate diplomacy.

Climate change-induced crises disproportionately affect developing nations and conflict-ridden regions, intensifying existing vulnerabilities. Addressing this issue at COP29, President Erdoğan linked environmental degradation with geopolitical instability, stressing that climate-related destruction should not go unpunished. He asserted, "This humanitarian and environmental disaster must be accounted for in international courts. Those responsible for this catastrophe should face justice at the highest level." (Erdoğan, 2024). This perspective aligns with the broader principle of climate justice, which argues that those contributing most to environmental destruction should bear the responsibility for mitigation and adaptation efforts. Erdoğan's remarks also reflect Türkiye's diplomatic stance on environmental accountability, positioning the country as an advocate for stronger legal frameworks in climate governance.

6.5 Conclusion and Recommendations

Modern diplomacy increasingly requires a multilayered and adaptive approach, as global challenges like climate change and migration transcend national borders. Climate migration, a growing crisis at the intersection of environmental change, security concerns, and economic instability, has become one of the most pressing issues in international diplomacy. Despite international efforts to address environmental concerns through treaties like the Kyoto Protocol and the Paris Agreement, migration remains underrepresented in global climate governance frameworks. The gap between environmental diplomacy and migration diplomacy highlights the urgent need for an integrated approach that aligns climate action with migration policies.

This study, employing a mixed-methods approach, has provided empirical insights into Türkiye's migration policies and climate diplomacy strategies. Through quantitative analysis of datasets from UNHCR, IOM, and the World Bank, the research has identified the increasing scale of climate-induced migration affecting Türkiye. Additionally, qualitative content analysis has revealed that while Türkiye actively engages in international migration diplomacy, its climate policies remain primarily reactive rather than preventive. Türkiye's dual role as both a transit and destination country underscores the diplomatic challenges it faces, particularly as migration pressures intensify due to drought, extreme weather events, and regional conflicts exacerbated by climate change.

Climate migration poses a complex diplomatic challenge, as it requires states to balance humanitarian responsibilities, national security concerns, and economic stability within a volatile global framework. Türkiye, given its strategic geopolitical location, finds itself at the center of this dilemma. While Türkiye has leveraged its position in negotiations with the EU, particularly through migration-related agreements, climate migration introduces new uncertainties that demand long-term policy adjustments. The study's findings emphasize that Türkiye must move beyond short-term agreements and develop a long-term, sustainable climate migration strategy.

To effectively address the challenges of climate migration, Türkiye and the international community must take several key diplomatic and policy measures. First, bridging the gap between climate diplomacy and migration diplomacy is essential. A unified framework should be established to integrate climate adaptation policies into migration governance, ensuring that climate-induced displacement is formally recognized in international agreements. Second, financial and technical support for climate adaptation must be strengthened. Türkiye must advocate for greater access to international climate finance to implement adaptation strategies, particularly in regions prone to forced displacement. Third, migration policies must be more inclusive. As a key transit country, Türkiye should prioritize social and economic integration programs for climate migrants to mitigate the risks of instability and economic marginalization. In addition, Türkiye should enhance its soft power in climate diplomacy. By engaging in public diplomacy efforts, Türkiye can shape international narratives, advocate for climate justice, and position itself as a leader in climate migration governance. Finally, there is an urgent need to establish a legal framework for climate migrants. International agreements must evolve to ensure that climate-displaced populations receive legal recognition and protection, preventing further humanitarian crises.

Ultimately, climate migration represents one of the most significant diplomatic tests of the twenty-first century. Türkiye, due to its strategic geographical location and extensive diplomatic experience in migration governance, is uniquely positioned to lead global discussions on this issue. However, without a coordinated and forward-thinking approach, climate migration could exacerbate geopolitical tensions, deepen global inequalities, and destabilize already fragile regions.

The world now stands at a pivotal moment. Nations must either take proactive diplomatic action to integrate climate migration into global governance structures or face an escalating crisis with no structured response mechanisms. Türkiye's role in shaping climate migration diplomacy will be critical in defining international policies that ensure fairness, resilience, and cooperation in addressing climate-induced displacement. Without strong leadership and global commitment, climate migration risks becoming an irreversible crisis that threatens international stability, economic development, and humanitarian security.

References

Abdurrahmanlı, E. (2021). Definition of diplomacy and types of diplomacy used between states. *Anadolu Akademi Sosyal Bilimler Dergisi*, *3*(3).

Alver, S. (2024). *Digital diplomacy: Twitter sentiment analysis of Donald Trump presidency* (Master's thesis, Yeditepe University, Institute of Social Sciences, Department of Political Science and International Relations).

An, N., Turp, M. T., & Kurnaz, L. (2021). İklim değişikliğine bağlı çevresel bozulmanın göç kararına etkisi: Genel bir bakış. *Ege Coğrafya Dergisi*, *30*(2), 383–403. https://doi.org/10.51800/ecd.932879

Arar, G. (2024). *Çevre Diplomasisi Dediğimiz...*, https://www.mfa.gov.tr/cevre-diplomasisi-dedigimiz___-.tr.mfa, Accessed: December 2024

Argalı, M. (2021). *Analysis of migration policy and migration diplomacy cases applied by Türkiye* (Master's thesis, National Defense University).

Arıkan, Y., & Özsoy, G. (2008). A'dan Z'ye iklim değişikliği başucu rehberi. Bölgesel Çevre Merkezi - REC Türkiye. Tuna Matbaacılık San. ve Tic. AŞ. https://rec.org.tr/wp-content

Avar, Y. (2016). *The development of inter-state relations and the history of diplomacy.* Tuiç Academy. https://www.tuicakademi.org/devletlerarasi-iliskilerin-gelisimi-diplomasi-tarihi/

Barratt, J. (1985). South African diplomacy at the UN. In G. R. Berridge & A. Jennings (Eds.), *Diplomacy at the UN* (pp. 191–203). Palgrave Macmillan.

Birleşmiş Milletler. (2015). Sürdürülebilir kalkınma amaçları: SKA 13 – İklim eylemi. United Nations. https://www.un.org/sustainabledevelopment/climate-action/

Bull, H. (1932). *The encyclopedia of diplomacy.* The Australian National University.

Çiçek, A., & Turan, A. (2023). Public diplomacy, migration and refugee politics: The case of Türkiye. *R&S – Research Studies Anatolia Journal*, 6(4), 584–613. https://dergipark.org.tr/rs

ÇŞB. (2018). *Birleşmiş Milletler iklim değişikliği çerçeve sözleşmesi.* Ministry of Environment and Urbanization. https://iklim.csb.gov.tr/birlesmis-milletler-iklim-degisikligi-cerceve-sozlesmesi-i-4362

Cull, N. J., & Rodríguez Jiménez, F. J. (2015). Introduction: Soft power, public diplomacy, and democratization. In *US public diplomacy and democratization in Spain* (pp. 1–15). Palgrave Macmillan. https://doi.org/10.1057/9781137461452_1

Directorate of Communications. (2022). *Kamu diplomasisi nedir?* Cumhurbaşkanlığı İletişim Başkanlığı Yayınları.

Directorate of Communications. (2022). *Public diplomacy in international relations.* Turkish Presidency of Communications.

Erdoğan, R. T. (2024). Speech at the COP29 World Leaders Climate Summit. Delivered at the United Nations Climate Change Conference (COP29), Baku, Azerbaijan. https://www.iletisim.gov.tr/turkce/haberler/detay/cumhurbaskani-erdogan-cop29-dunya-liderleri-iklim-zirvesinde-konustu

IOM. (2024). *Climate change and future human mobility*: Evidence summary. https://emergencymanual.iom.int/sites/g/files/tmzbdl1956/files/202303/iom_global_data_institute_thematic_brief_1_evidence_summary_on_climate_change_and_the_future_of_human_mobility.pdf

İstanbul Planlama Ajansı. (2024). İklim krizini dezavantajlı gruplar üzerinden okumak: Türkiye ve İstanbul'da iklim adaletine yönelik bir değerlendirme. https://ipa.

istanbul/wpcontent/uploads/2024/08/KENT_GUNDEMI-IklimKriziniDezavantajl iGruplarUzerindenOkumak.pdf

IPCC-AR6. (2023). *Climate change 2023: Synthesis report, summary for policymakers.* A report of the Intergovernmental Panel on Climate Change, Geneva, Switzerland.

Kentli. (2022). İklim değişikliği ve göç. Kentli Dergisi, *45*, 41–51. Türkiye Sağlıklı Kentler Birliği, Bursa. https://www.skb.gov.tr/wp-content/uploads/2022/07/Kentli-45.-Sayi.pdf

Koç, S. N. (2024). *Analysis of Türkiye-Bulgaria relations in the context of migration diplomacy (1923–1989)* (Master's thesis, Bursa Uludağ University).

MFA. (2020a). *BM iklim değişikliği çerçeve sözleşmesi.* Ministry of Foreign Affairs of the Republic of Turkey. http://www.mfa.gov.tr/bm-iklim-degisikligi-cerceve-sozlesmesi.tr.mfa

Ökten Sipahioğlu, B. (2022). Küresel iklim değişikliğinin bir sonucu olarak iklim göçü. *Akademik Hassasiyetler / The Academic Elegance*, *9*(20), 543–565.

UN Women. (2022). *İklim değişikliği bağlamında kadınlara ve kız çocuklarına yönelik şiddetle mücadele.*

UNHCR. (2024). *Türkiye'deki mülteciler ve sığınmacılar.* UNHCR. https://www.unhcr.org/tr/turkiyedeki-multeciler-ve-siginmacilar

7

Local Actors in Climate Diplomacy

Case Studies

Brígida Brito

7.1 Introduction

Over time, the European Union (EU) has become one of the most important players in international environmental policy, particularly on the climate issue. The concern over environmental problems has led to the adoption of a proactive approach in agreements negotiation, the creation of ecological networks, and the promotion of targeted measures to achieve global goals. In addition to the interest in world environmental conditions, particularly with regard to extreme weather events and their impacts, in recent decades the EU has mobilized financial and technical efforts to facilitate support for developing countries in this area of intervention.

Although European cooperation with the African, Caribbean and Pacific (ACP) countries began in the mid-1970s with the Lomé Conventions, which were reconverted in 2000 with the Cotonou Agreements (Brito, 2024, p. 607–614), the greatest involvement took place precisely during the twenty-first century. Initially, cooperation between the EU and the ACP countries focused on dispersed thematic areas of intervention, such as international trade, agricultural cooperation and food security, environmental preservation, industrial development, and services, among others. However, in recent years, the growing concern with adaptation to climate change and, depending on the case, mitigation has gained significance. This focus is evident in the guidelines for cooperation between the EU and the ACP countries, with implications for financial support.

The United Nations Framework Conventions on Climate Change (UNFCCC) have contributed to the EU's increasing awareness of the issue, both domestically through the creation of a multitude of programs and projects, and internationally through climate cooperation. The EU has taken a leading role in climate policy with the ambition of carbon neutrality by 2050. In this regard, it has firsthand introduced innovation in actions. On the other hand, a global negotiation was promoted which is materialized in the national plans adopted by the member states. But also in the international influence exerted, particularly in territories affected by the impacts of extreme weather events and where vulnerability is recognized. It is in this context that climate diplomacy is highlighted and has been strengthened, seeking negotiation with partner states, but also facilitating collaboration and creating conditions for effective cooperation in the environmental and climate area.

This analysis seeks to answer the following question: how do local actors in ACP countries participate in climate diplomacy? From a methodological point of view, a case study was chosen to highlight the capacity to mobilize resources, the willingness to negotiate conditions within the framework of the annual Climate Summits, and the involvement in establishing international partnerships.

7.2 The Framework for Climate Diplomacy

Climate diplomacy is defined as the new diplomacy (Nadir et al., 2020, p. 316) in the sense that it is a new foreign policy instrument adjusted to environmental issues. From a methodological point of view, it focuses on international negotiation based on debates and proposals for action that are close to the concept of soft power developed by Joseph Nye (2004). This concept advocates the involvement of local actors, including non-state actors.

Over time, the terminology adopted has evolved with a diversity of concepts, including expressions such as green diplomacy, also known as ecodiplomacy, energy diplomacy, climate diplomacy, and water diplomacy (Nadir et al., 2020, p. 320). These concepts are naturally linked to the problems that promote them. Thus, problems related to the management and maintenance of ecosystems, the (un)availability of energy sources, greenhouse gas emissions, the management of water resources, including shared watershed basins, and extreme weather events can be mentioned. The soft power framework comes from the understanding that if the diplomatic dimension is not considered, the different perspectives on environmental issues would not be valued. All these problems have a strong impact from a social point of view, affecting people, but also the economic sphere, contributing to the worsening of crises, and the environmental area, resulting in losses of resources and transformation of ecosystems. All in all, there are added risks to sustainability.

In the international context, the environment is now clearly defined as a public commodity (Nadir et al., 2020, p. 322), requiring the definition of a specific agenda that contributes to environmental preservation and conservation of species with endangered status or endemics. However, the focus is particularly on the well-being of local populations, especially those living in poverty. In this sense, the diplomatic agenda favors interaction between a multiplicity of actors. We can consider the bilateral ones, which are defined as conventional, the multilateral ones, which are strengthened given the suitability associated with the absence of self-interest, and the emerging civil society organizations that work according to proximity methodologies to the problems and affected communities.

The role of environmental diplomacy has been reinforced by the creation of specialized agencies, including the United Nations Environment Programme (UNEP), the Convention on Biological Diversity (CBD), the Intergovernmental Panel on Climate Change (IPCC), and the United Nations Framework Convention on Climate Change (UNFCCC). Within the framework of this enhancement, concerns about safety and environmental governance have become preponderant (Nadir et al., 2020, pp. 325–332), considering that these factors ultimately refer to the human dimension. Faced with the multiplication of extreme climatic events, particularly in sensitive geographical areas, and the progressive intensity of their impacts, climate has come to be seen as the priority of environmental negotiation and therefore the focus of diplomacy. Through the attempt to control the anthropogenic factors that intensify them, climate change, and more specifically extreme weather events, have become the main international challenge that has been mobilizing efforts within the framework of climate diplomacy.

Although the new model of international relations in the environmental area can be traced back to the early 1970s, the emphasis on climate diplomacy took place at the turn of the millennium. On the one hand, it resulted from the intervention of multilateral organizations and, on the other, from the growth of social movements which, by formalizing themselves, made it possible to set up non-governmental environmental organizations. Although at an early stage there was no consensus, the climate issue received attention from all organizations, including public, private, civil society, national, and international. The beginning of the twenty-first century was marked by the worsening of the climate issue and, consequently, its impacts on humans, which highlights a growing global concern.

One of the most charismatic examples of environmental diplomacy in the area of climate is the United Nations Conferences of the Parties, known as the Climate COPs (Bremberg, 2024). In this sense, COP21 and the Paris Declaration can be considered one of the milestones of climate diplomacy, even if it didn't achieve all the proposed goals. Portugal's Foreign Minister at the time wrote, "The negotiation of the Agreement was lengthy and complex, but it was successfully concluded. The same success was achieved in the ratification process'[1] (Silva, 2020, p. 1). This is possibly one of the most successful cases in this area. However, in the last ten years (2015–25) it has suffered from advances and setbacks as a result of some of the largest emitters of greenhouse gases, which represent the greatest contribution as an anthropogenic factor to the worsening climate situation, having intermittent and inconsistent participation in agreements and the implementation of action measures. As an example, the cases of the United States of America, China, and India confirm this.

The concept of climate diplomacy is accompanied by another, that of climate action. This conceptual combination assumes the understanding that stabilizing the climate and its effects, which affect billions of people around the world, is a global responsibility and so it must be shared. This assumption is also realized on the basis of the principle of proportionality, that is, the countries that emit the most are the ones that should be committed to a greater reduction in emissions. This is a global, regional, national, and local challenge that requires concerted action by all stakeholders in favor of a common goal, that of sustainability. It is in this sense that it is argued that "Climate action is one of the fundamental areas of multilateral cooperation; it is everyone's problem, which only together we can resolve, or even alleviate. It requires concertation and compromise, a very different attitude and language from the fracture between two rival political, economic and technological orders that has paralysed the world in recent years"[2] (Silva, 2020, p. 3).

In the European context, on July 3, 2018, the concept was formalized through the European Parliament Resolution on Climate Diplomacy (Official Journal of the European Union, 2020). Climate diplomacy is defined on the basis of effective cooperation with partner states, albeit with an obvious multilateral focus. Diplomacy and cooperation are perceived as two sides of the same coin, working together to achieve the challenge of climate stabilization. Afterwards, the European Council adopted the conclusions on Climate and Energy Diplomacy (Council of the European Union, 2021), centered on achieving the European Green Deal. The possibility of actors other than centralized state entities being able to develop action plans has become more of a reality than an opportunity. This includes decentralized entities such as municipalities and cities, which are recognized for their active engagement in the context of paradiplomacy (Sassen, 2012, p. 102; Acuto, 2015; Curto et al., 2014 and Moita, 2017). In addition to these, local non-state organizations are also strengthened because of their capacity to work in the field and because they implement proximity-based methodologies.

7.3 The Importance of Multilateral Actors in Environmental Paradiplomacy

Paradiplomacy is an expression that is centered on a practice of parallel negotiation and cooperation promoted by non-state actors, or predominantly non-state actors, which can also be defined as sub-state actors. In the various areas of intervention, it is common to mention multilevel activities beyond the state, which in most cases result from processes of debureaucratization (Cornago, 2010; Curto et al, 2014; Tavares, 2016). It's a concept that reflects the evolution of the globalization process and the valuation of international partnerships and networks that are formalized with the involvement of non-state actors.

According to Neves (2010, p. 25),

> it shows that external action will be more and more a multi-actor, multidimensional process where public, private and third sector actors have to engage and combine their different skills in the context of long term partnerships. Knowledge networks involving coordination and cooperation between governments, business, NGOs, academia, trade unions becomes paramount for effective external action not only in terms of implementation but also in terms of policy conception.

The belief that non-state actors, particularly local ones, can take on the negotiation of actions, projects and programs through their own channels has acquired relevance over time and is now a reality.

In addition to other relevant areas, paradiplomacy is rediscovered in environmental studies, particularly in the identification of problems, the definition of partnerships, and the design of action methodologies to minimize the effects or eventually resolve risk factors. The environment is one of the favored areas of paradiplomacy, which creates space for the emergence of climate diplomacy, also known as "new diplomacy," following the terminology used by Mohammed Nadir (2020, p. 316).

At global level and over time, the actions of the state in the regulation of environmental problems have proved to be insufficient and ineffective, as the environment has often taken second place to national priorities. In fact, the environmental issue, and even more so the climate problem, cannot follow a unilateral or one-dimensional approach. This is a global and systemic problem that requires an integrated intervention, just like the factors that degrade or worsen environmental conditions, which are known as anthropogenic, following the approach of Paul Crutzen (2002) and Will Steffen (2015).

Nowadays, environmental risks are prioritized in connection with climate variability and its effects, which in many circumstances become unsustainable because they make human life unviable. The acceleration of climate change is directly related to the issue of energy, and this is equated with security in all its many meanings. Several multilateral organizations are calling for a joint approach to ensure that this triple problem is met, which can be simply summarized as the link between climate, energy, and security.

Within the framework of the UNFCCC, the Green Climate Fund (GCF) was created, a financial instrument that enables cooperation with developing states, and that was agreed by consensus between 194 countries. The aim of the measure is to identify and establish alternative, global climate regulation models. The Green Climate Fund (GCF, n.d.) is a multilateral mechanism for the allocation of resources to finance or support low-emission projects and programs. These tend to be defined as essential contributions to the energy transition in countries that lack applied research and development initiatives, known as research and development (R&D) projects. The GCF gives special prominence to African countries and Small Island Developing States (SIDS). In this context, in 2024 there have been several projects promoted by climate diplomacy through the GCF in 133 developing countries,[3] which highlights the importance of multilateral climate diplomacy mechanisms. Adaptation projects aim to increase the resilience of local people and vulnerable ecosystems to the impacts of extreme weather events. On the other hand, mitigation projects are designed to reduce emissions by reinforcing the function of natural and artificial carbon sinks.

In order to continue supporting the least developed countries, at COP28, the Loss and Damage Fund[4] with a budget amount of US$100 billion that was proposed at the previous COP was approved(Euronews, 2023). This is a compensation mechanism centered on what can be defined as emergency aid, since its purpose is to support people who are affected by the impacts of extreme weather events. In fact, this compensation is given to the least developed states, which are responsible for managing the support, and not to the communities. Although this represents a step forward in climate diplomacy, it also raises doubts about the operationality of the funds for the resolution of emerging problems.

In the EU's sphere of intervention, partnerships and commitments are being proposed in order to promote the resolution of the challenges that link climate change to environmental degradation, but also to peace and security. Climate cooperation can be seen as part of the so-called green diplomacy, which clearly searches for solutions and alternatives to the systematic use of fossil fuels. Considering this awareness and the adjustment of actions to the commitments made when the Paris Agreement was ratified in 2016, some of the cooperation tools set up by the European Union for African, Caribbean and Pacific countries provide for the pursuit of projects that combine the energy sector with waste treatment and climate. The allocation of EU funds is directed toward the implementation of initiatives that contribute to adaptation and, in some cases, mitigation. In this context, a mixed approach is foreseen among the different stakeholders, favoring the multilateral dimension of funding and non-state or subnational actors in the implementation of actions on the ground. These solutions are about achieving extensive and global results in a fair and equitable way, prioritizing stakeholders who otherwise wouldn't be able to access the funding.

The focus of multilateral organizations in regions such as the African continent seems obvious considering the institutional weaknesses at the national level that, in themselves, limit the adoption of innovative models, both technical and managerial. Climate regulation through energy, waste collection and treatment, or habitual consumption is, in many cases, made impossible by the state itself, which invests very little in these areas, setting other priorities. "International concern has centred on the fragility of the climate regulation models adopted by most African countries, which reveal vulnerability due to the increase of risks, event after event, affecting people who clearly live in poverty"[5] (Brito, 2023, p. 70).

National responses from the state show weaknesses, both because of the unavailability of financial resources and because of the lack of technical knowledge that makes it impossible to predict and safeguard or correct after an event with social, economic, and environmental impacts. "In the African context, the reduced, often non-existent, capacity for adaptation and mitigation is evident, making it difficult to achieve the objectives assumed at international level and clarified by Climate Action as described in the thirteenth Sustainable Development Goal."[6] (Brito, 2023, p. 70).

After 2016, and keeping in mind the challenging targets that had been set at COP21 in Paris, new global commitments were agreed favorable to the creation of cooperation mechanisms with multi-level intervention, which can also be defined within the framework of climate diplomacy. In this regard, new instruments have been created and those previously established have been strengthened, aimed both at financing adaptation and mitigation initiatives and at technical cooperation to make the energy transition viable.

One of the current focuses of the European Union's foreign policy with its partners, including African countries, is precisely the energy transition. One example of the mechanisms created is the Global Gateway Strategy, which seeks to create the right conditions for a sustainable approach based on resilience at various levels, including climate, health, and competitiveness. Within this framework, the Europe-Africa Investment Package was created, making 150 billion euros available to promote energy transition in an accelerated manner, among other privileged areas (European Commission, n.d.).

One of the priority areas of intervention is the implementation of renewable energy in Africa, as can be seen in the case of the *Scaling Up Renewables in Africa* project (European Commission, n.d.).

There is now a general awareness that the African continent, mostly in rural areas, but also in the cities, lives in the shadows, in the dark, without lighting or with insufficient street lights to meet the daily needs of the population. This European cooperation project with African countries has the merit of combining energy accessibility with the fight against poverty, identifying a cooperation cluster that guarantees equal justice. It is estimated that on average, about 600 million people in Africa suffer from energy shortages. However, although it may seem contradictory, it is estimated that the continent has 60 percent of the world's solar energy sources, which are not used in favor of development or the well-being of local communities. According to the European Commission (n.d.), foreign investment has not capitalized on the potential of clean, green, or alternative energy sources, since it only accounts for 3 percent of global investment in the sector worldwide.

The analyzes on the subject suggest ambitious targets for decarbonization by 2050 and, to a large extent, the proposals for action that emerge from the COPs, and adopted by different multilateral agencies, are criticized by environmentalists, experts, and academics. The main reasons for this criticism are centered on the understanding, on the one hand, that the excessively ambitious targets are not achievable, particularly in poor countries where the path is long and time-consuming. The 2030 deadline for the Sustainable Development Goals, in particular, the SDG13 relating to climate action, or the 2050 deadline relating to the expectation of European decarbonization, is not suitable for the African continent. The dependence on fossil energy resources is still a reality, despite the fact that access to energy is not generalized to the entire population. This observation often gives rise to interpretations about the effectiveness of the democratic model, including in countries that produce and export oil or crude oil, showing signs of a lack of equity and societies devoid of socio-environmental justice. On the other hand, regulating the climate and its adverse effects will only be possible in the long term if all the world's countries fulfill their commitments. Given that a large part of climate diplomacy is not regulated by mandatory principles (Brito, 2023, p. 75), in line with the assumptions of soft power, the results achieved in poor countries such as African ones are always more fragile than initially expected.

7.4 Empowering Local Actors for Climate Action: The case of São Tomé and Príncipe

There are many cases that can be analyzed as climate diplomacy practices in Africa, supported by multilateral agencies and involving local actors. These are processes of paradiplomacy and soft power that do not prioritize state institutions, even though they may involve them.

One of these examples is centered on the work of The Global Environment Facility (GEF), a financial mechanism[7] created in 1992 as part of the United Nations Conference on Environment and Development, commonly known as the Rio Summit or the Earth Summit. As a proposal resulting from the Earth Summit meetings, environmental justice is emphasized as an international priority and therefore a global challenge. The GEF's objective is to promote balanced and fair environmental conditions worldwide, prioritizing developing countries. Over time, the GEF has become part of six funds: the Global Environment Facility Trust Fund; the Global Biodiversity Framework Fund; the Nagoya Protocol Implementation Fund; the least developed countries Fund; the Capacity Building Initiative for the Transparency Trust Fund; and the Special Climate Change Fund.

The GEF is composed of 186 members, supporting 1,700 environmental projects and programs in 149 countries (The GEF, n.d.). More than a third of the projects and programs financed are implemented in Small Island Developing States (SIDS) and in least developed countries. The GEF's priority sectoral areas for action in different countries are conservation and biodiversity (36 percent of funding), climate change mitigation projects and programs combined with waste treatment (31 percent), soil degradation (12 percent), international water management (11 percent), and corporate programs (7 percent), while 4 percent goes to administrative costs (The GEF, n.d.).

Given that the GEF supports SIDS and Least Developed Countries, São Tomé and Príncipe is a relevant case for analysis. It is an African micro-state (Brito, 2021), located in the Central African subregion, with three main geographic reference elements: the Atlantic Ocean; the Gulf of Guinea; and the Equator (Latitude 0°04N). Being an island country, it is an archipelagic one, made up of two principal islands (the island of São Tomé and the island of Príncipe) and a group of mostly uninhabited islets. The soil is of volcanic origin, which gives altitude and irregularity to the relief, accentuated by a hydrographic network[8] extensive and radial in outline with sources in the interior of the two islands and the estuary of the rivers flowing into the sea. The two islands are characterized by abundant watercourses and abrupt cliffs. The country is dotted with many peaks, with Pico de São Tomé being the highest at 2,024 meters (The World Bank, 2023).

From an administrative point of view, the 1,001 square kilometers of territory[9] are divided into seven districts,[10] six on the island of São Tomé and one on the island of Príncipe, all confluent with the sea. The coastline is 260 kilometers long and the total area of the Exclusive Economic Zone is 160 times the territorial area, that is, 160,000 square kilometers.

Given its geographical location and the influence of the equator, the climate is characteristically tropical and humid, with two well-defined seasons: the rainy season and the gravana, or dry season. Although the entire territory is influenced by high levels of humidity throughout the year, reaching 90 percent, the districts with the highest rainfall are located in the South and West. According to data from the 2024 general population census (INE STP, n.d.), although small, the national population (N = 209,607 inhabitants) is predominantly urban (67.7 percent), exerting pressure on the spaces and natural resources on which communities depend. The economy is centered on the primary sector associated with the tertiary sector, with informal commerce and services playing a major role.

Since the period of political independence,[11] the country has been confronted with a series of structural problems associated with socioeconomic dependence on nature, which results in unplanned and sometimes excessive use of natural resources, contributing to their depletion. External dependence marks the national macroeconomy, as the country has been unable to find internal financing instruments to revitalize the productive sector and innovate. But at the same time, as a SIDS, it suffers from what is commonly known as the Small Island State syndrome (Brito, 2005), which naturally highlights its insularity, small size, distance from international centers, isolation, and vulnerability.

From an environmental—or socio-environmental—point of view, the country has attracted worldwide attention not only because it is an internationally dependent SIDS from a financial and technical point of view, but mainly because it fulfills a set of conditions that make it a paradigmatic case. As a systematization, we can list some of the aspects that make São Tomé and Príncipe a case study of great importance in the context of SIDS supported by multilateral mechanisms and following the criteria of environmental diplomacy (Government of STP, 2021, n.d.):

a. One third of the national territory is legally classified as a National Park, which gives it national protection status, the Obô National Park present on the two main islands;
b. With regard to anthropogenic factors influencing climate change, the country is defined as a double carbon sink, both through forestry and the ocean;
c. Although it does not represent an anthropogenic threat to the world, on the contrary, it is a country that regulates climate change. It participates in the Climate COP and other meetings, ratifying agreements and negotiating treaties to implement measures;
d. In 2012, the entire island of Príncipe was classified by UNESCO as a Biosphere Reserve, which implies that the country is part of the Man and the Biosphere program and is promoting actions to preserve ecosystems and to conserve species;
e. Both islands are endowed with biodiversity of fauna and flora in forest, river, coastal, and marine ecosystems, including endemic species that are naturally vulnerable due to their status;

f) Some of the species are threatened and have been assigned endangered status by the International Union for Conservation of Nature (IUCN) through its Red List;

g. The local population, both on the island of São Tomé and on the island of Príncipe, maintains primary and basic models of exploitation, production, and use of resources, sometimes deriving from a recollective practice, and other times following a family-based production model without showing systematic and regular ability to make a profit;

h. The microeconomic logic is still oriented toward direct consumption and survival, possibly with the sale of surpluses, which guarantees the acquisition of income, albeit in a model marked by informality;

i. Since the colonial period, and continuing to the present day, the occupation of physical space has been defined according to two major geographical and productive areas. First, the coastline, mostly occupied by fishing families,[12] some of whom continue to practice itinerant nomadism, which means more or less prolonged absences before returning home. Second, the roças, old agricultural production units that followed the logic of the production cycles associated with the monoculture of sugar cane, coffee, and cocoa. After independence, these were occupied by families of Cape Verdean origin who established themselves on the land where their ancestors produced export crops but stopped making a profit from them;

j. As this is a SIDS located on the Gulf of Guinea, influenced by the equator, high rainfall rates define the conventional climatological pattern, which means intense rainfall associated with warm temperatures;

k. The effects of climate change in this archipelagic territory are many and varied, namely: a variation in the climate pattern in terms of a decrease in the intensity of rainfall, changes in the two conventional seasons, and a worsening of drought periods; a rise in sea level affecting fishing communities settled on the coastline and near the beaches and in low-lying localities; an increase in the flow of watercourses such as rivers and streams, which in high-lying localities risks landslides and flooding affecting communities settled in valleys;

l. The risk of losing biodiversity, even if only occasionally, is a reality, since environmental variations, such as climate change, also force species of fauna and flora into a process of adaptation;

m. The obvious development of the tourism sector, in particular the nature, ecological, and ecotourism segments, given the increase in international demand for trekking and watching activities (birdwatching, whalewatching, turtlewatching etc.).

This wide range of factors is not exhaustive and is intended to exemplify the interest of multilateral entities in supporting socio-environmental actions, projects and programs to strengthen resilience in São Tomé and Príncipe. In this context, the United Nations, through its agencies in the country, such as the United Nations Development Programme (UNDP), has created a set of support instruments to consolidate the environmental SDGs, namely SDG 13 on Climate Action, SDG 14 on Marine Life, and SDG 15 on Terrestrial Life (Brito, 2021).

Within the framework of its financial competencies, the GEF has been supporting São Tomé and Príncipe over time in different types of actions, projects and programs, stimulating the involvement of different stakeholders, in particular civil society organizations that work directly with communities and in close proximity to the emerging problems. Following a brief chronological guideline, which is not exhaustive but clarifying, some of these supports are:

a. In 2014, jointly, the GEF and the International Fund for Agricultural Development (IFAD) funded an integrated program called the Participatory Program for the Development of Family Farming and Artisanal Fishing (PAPAFPA). As part of the activities carried out, the GEF funded the production and publication of the Environmental Education Manual, São Tomé and Príncipe (Pinto and Lima, 2014), which was implemented to support the Environmental Education curriculum enrichment area for secondary schools. The project involved organizing and carrying out training sessions with teachers and trainers and resulted in a partnership cooperation action implemented by two civil society organizations, Mar, Ambiente e Pesca Artesanal (MARAPA) from São Tomé and Príncipe and the Associação Portuguesa de Educação Ambiental (ASPEA) from Portugal. In the wide range of topics analyzed in the Manual and in the training sessions, climate and environmental risks were highlighted in order to raise awareness among the different groups that were involved. The training contributed to raising awareness for action, creating a critical mindset about conventional socio-cultural practices and representations that result in environmental aggression with a negative impact.

b. In 2018, the GEF created conditions to support the resilience of coastal communities through the West African Coastal Areas (WACA) Resilience Investment Project, which involved an investment of US$210 million in six countries with significant coastline settlements, including São Tomé and Príncipe. According to the World Bank's Vice-President for the Africa Region, "The WACA Resilience Investment Project is a collective response to the urgent need to address coastal degradation in a regional, integrated manner. It is an opportunity to strengthen the resilience of West Africa's communities and transform their livelihoods" (The GEF, s.d.).
c. With the purpose of creating conditions for the energy transition, the GEF supported the Renewable Energy and Energy Efficiency Project in the electricity sector of São Tomé and Príncipe, implemented by UNIDO, the United Nations Agency for Industrial Development, and the Lusophone Renewable Energy Association. The report that was presented in 2020 indicates different opportunities, including for women. At the same time, and in a complementary way, UNIDO has supported the implementation of the Strategy for Small Island Developing States, which involves capacity building through webinars and technical training, particularly in the area of renewable energy. On the other hand, there is the Strategic Program for the Promotion of Investments in Renewable Energy and Energy Efficiency in the Energy Sector of São Tomé and Príncipe, implemented by UNIDO. This program is entirely framed by the strategic plan called São Tomé and Príncipe 2030: The Country We Must Build.

In any of the above cases, the involvement of civil society groups is taken into account, either as stakeholders and partners in the projects or through training and capacity building in order to generate or strengthen future opportunities. Empowerment is always present, taking into account that the stakeholders become aware of the risks, but also, and above all, of the multiplicity of possible paths toward a more equitable and resilient socio-environmental relationship.

7.5 Conclusion

Although the models of environmental and climate diplomacy are multiple and diverse, the institutional framework of multilateral organizations represents an asset for the least developed countries, or those suffering from vulnerabilities, and their stakeholders. Multilateral instruments represent security mechanisms, supporting decision-making, guaranteeing funding for targeted projects, facilitating technical support, and reinforcing positions in international agreements.

While climate diplomacy cannot, and should not, be seen as a miracle solution to all socio-environmental problems, it is an instrument of soft power centered on negotiation and acts as a facilitator in promoting development with environmental enhancement.

As a point of reflection,

> Africa represents today only 2% of the global emissions, due to the low level of energy consumption per capita. This consumption will have to increase. If we want Africans—who are growing in numbers—to have the same level of quality of life, they will have to increase their energy consumption and, at the same time, not increase the CO2 emissions. This will require a gigantic effort and we will have to be standing together with our African partners. Climate will be a key part of our African partnership. (Borrell, 2021)

In the case of vulnerable countries, particularly Small Island Developing States, climate diplomacy in the form of paradiplomacy has achieved results. The analyzed case of São Tomé and Príncipe is an example of the various actions undertaken using different types of actors—multilateral, civil society, and institutional participation—in the search for solutions to the sustainability challenge.

Notes

1 Translated by the author from the original in Portuguese.

2 Translated by the author from the original in Portuguese.

3 Afghanistan, Albania, Algeria, Angola, Antigua and Barbuda, Argentina, Armenia, Azerbaijan, Bahamas, Bahrain, Bangladesh, Barbados, Belize, Benin, Bhutan, Bolivia, Bosnia and Herzegovina, Botswana, Brazil, Burkina Faso, Burundi, Cabo Verde, Cambodia, Cameroon, Central African Republic, Chad, Chile, China, Colombia, Comoros, Congo, Cook Islands, Costa Rica, Cuba, Côte d'Ivoire, Democratic People's Republic of Korea, Democratic Republic of the Congo, Djibouti, Dominica, Dominican Republic, Ecuador, Egypt, El Salvador, Equatorial Guinea, Eritrea, Eswatini, Ethiopia, Fiji, Gabon, Gambia, Georgia, Ghana, Grenada, Guatemala, Guinea, Guinea-Bissau, Guyana, Haiti, Honduras, India, Indonesia,

Iran, Iraq, Jamaica, Jordan, Kazakhstan, Kenya, Kiribati, Kuwait, Kyrgyzstan, Lao People's Democratic Republic, Lebanon, Lesotho, Liberia, Libya, Madagascar, Malawi, Malaysia, Maldives, Mali, Marshall Islands, Mauritania, Mauritius, Mexico, Micronesia, Moldova, Mongolia, Montenegro, Morocco, Mozambique, Myanmar, Namibia, Nauru, Nepal, Nicaragua, Niger, Nigeria, Niue, North Macedonia, Oman, Pakistan, Palau, Panama, Papua New Guinea, Paraguay, Peru, Philippines, Republic of Korea, Rwanda, Saint Kitts and Nevis, Saint Lucia, Saint Vincent and the Grenadines, Samoa, Sao Tome and Principe, Saudi Arabia, Senegal, Serbia, Seychelles, Sierra Leone, Singapore, Solomon Islands, Somalia, South Africa, South Sudan, Sri Lanka, State of Palestine, Sudan, Suriname, Syrian Arab Republic, Tajikistan, Tanzania, Thailand, Timor-Leste, Togo, Tonga, Trinidad and Tobago, Tunisia, Turkmenistan, Tuvalu, Uganda, Uruguay, Uzbekistan, Vanuatu, Venezuela, Vietnam, Yemen, Zambia, Zimbabwe.

4 Germany and the United Arab Emirates offered $100 million each, and the European Union offered $225 million, with contributions from other states.

5 Translated by the author from the original in Portuguese.

6 Translated by the author from the original in Portuguese.

7 The GEF supports six international conventions, namely the Convention on Biological Diversity, Minamata Convention on Mercury, Stockholm Convention on Persistent Organic Pollutants, Biodiversity Beyond National Jurisdiction Agreement, UN Convention to Combat Desertification, and UN Framework Convention on Climate Change.

8 The country has 233 watercourses and 116 river basins.

9 Officially, the territory of São Tomé is defined by the limit of 1,001 square kilometers. However, the latest topographical studies indicate a loss of territory as a result of climate change and rising sea levels, and the real size of the country is considered to be 960 square kilometers. (Vila Nova, 2021).

10 On the island of São Tomé, the districts of Água Grande, where the country's capital is located, Mé-Zoxi, Lobata, Cantagalo, Caué, and Lembá; on the island of Príncipe, the district of Pagué.

11 The Democratic Republic of São Tomé and Príncipe became independent on July 12, 1975, after a long period of Portuguese colonization lasting more than 500 years (1470–1975).

12 It is common to associate fishing communities with *Angolares* and the occupation of the roças by descendants of slaves of Cape Verdean.

References

Acuto, M. (2015). City diplomacy. In P. Kerr et al. (Ed.), *SAGE handbook of diplomacy*. Sage.

Borrell, Josep (2021). *Glasgow: o papel decisivo da diplomacia climática da EU.* Available at https://www.eeas.europa.eu/eeas/glasgow-o-papel-decisivo-da-diplomacia-clim%C3%A1tica-da-ue_pti [Accessed on january 19, 2025].

Bremberg, N., & Michalski, A. (2024). The European Union Climate Diplomacy: Evolving practices in a changing geopolitical context. *The Hague Journal of Diplomacy*, *19*(3), 506–535. https://doi.org/10.1163/1871191x-bja10188.

Brito B. (2021). Conserving Marine Life in Sao Tome and Principe: Concerted Actions with Agenda 2030. In W. Leal Filho, A. M. Azul, L. Brandli, A. Lange Salvia, & T. Wall (Eds.), *Life below water*. Encyclopedia of the UN Sustainable Development Goals. Springer. https://doi.org/10.1007/978-3-319-71064-8_152-1

Brito, B. (2024). Europe-Africa cooperation: From the idea to partnerships to promote sustainable development. In Y. Li, F. J. B. S. Leandro, J. Tavares da Silva, & C. Rodrigues (Eds.), *The Palgrave handbook on China-Europe-Africa relations*. Palgrave Macmillan. https://doi.org/10.1007/978-981-97-5640-7_29

Brito, Brígida (2005). *O Desenvolvimento para além do Petróleo, o exemplo do Turismo em São Tomé e Príncipe*. Universidade de Aveiro.

Brito, Brígida (2023). Da cooperação climática aos desafios que persistem na África Subsariana. In *Janus 2023, Anuário de Relações Exteriores*. Disponível em https://observare.autonoma.pt/anuario/wp-content/uploads/sites/5/2023/09/Da-cooperacao-climatica-aos-desafios-que-persistem-na-Africa-Subsariana.pdf [Accessed on September 20, 2024].

Comissão Europeia (s.d.). Estratégia Global Gateway. Disponível em. https://commission.europa.eu/strategy-and-policy/priorities-2019-2024/stronger-europe-world/global-gateway_pt

Cornago, Noé (2010). On the normalization of sub-state diplomacy. *The Hague Journal of Diplomacy*, *5*(1/2), 11–36. https://doi.org/10.1163/187119110X12574289877281

Council of the European Union. (2021). *Council conclusions on climate and energy Diplomacy – Delivering on the external dimension of the European Green Deal.* General Secretariat of the Council.

Crutzen, P. (2002). Geology of mankind. *Nature*, 415, 23. Available at https://www.nature.com/articles/415023a.pdf. [Accessed on January 15, 2024].

Curto, H. S.; Moita., L.; Brito, B.; Quintas, C.; Galito, M. S. (2014). Cities and Regions: paradiplomacy in Portugal. Notes and Reflections, *JANUS.NET e-journal of International Relations*, Vol. 5, N. 2, November 2014–April 2015.

Euronews (2023). *Decisão histórica na COP 28 com a criação de fundo de "perdas e danos" relativo ao clima.* 1 de dezembro, Available at https://pt.euronews.com/ [Accessed on January 5, 2025].

European Commission (n.d.). Scaling up Renewables in Africa. Available at https://commission.europa.eu/strategy-and-policy/priorities-2019-2024/stronger-europe-world/global-gateway/energy-africa_en [Accessed on january 19, 2025].

GCF (n.d.) Available at https://www.greenclimate.fund/. [Accessed on january 19, 2025].

Governo de STP (n.d) (2021). *Plano de Acção Tecnológico para a Transferência das Tecnologias de Adaptação às Mudanças Climáticas no sector de Zonas Costeiras*. São Tomé e Príncipe.

Governo de STP (n.d.). Planos de Ordenamento da Orla Costeira. São Tomé e Príncipe.

INE STP (n.d.). Available at https://ine.st/. [Accessed on january 10, 2025].

Moita, Luís (Coord.) (2017). A internacionalização de Lisboa: Paradiplomacia de uma cidade. Lisboa: UAL, OBSERVARE. Available at http://hdl.handle.net/11144/3006 [Accessed on January 5, 2025].

Nadir, Mohammed, Tybusch, Jerônimo Siqueira, Araújo, Luiz Ernani Bonesso (2020). A diplomacia Ambiental no Sul Global: Um olhar sobre África e América Latina. In *Justiça do Direito*, v. 34, n. 1, p. 314–363.

Neves, Miguel Santos (2010). "Paradiplomacia, Regiões do Conhecimento e a consolidação do «Soft Power»." *JANUS.NET e-journal of International Relations*, N.º 1, Autumn 2010.

Nye, Joseph (2004). *Soft power. The means to success in world politics*. PublicAffairs.

Official Journal of the European Union (2020). Resolução do Parlamento Europeu, de 3 de julho de 2018, sobre Diplomacia climática (2017/2272 INI), (2020/C 118/04).

Pinto, J., & Lima, João Pessoa (coord) (2014). Manual de Educação Ambiental, São Tomé e Príncipe. ASPEA, MARAPA.

Sassen, S. (2012). Cities. In *the wiley-blackwell encyclopedia of globalization*. https://doi.org/10.1002/9780470670590.wbeog069 .

Silva, Augusto Santos (2020). Os desafios da ação climática para a diplomacia e o multilateralismo. In Agência Portuguesa do Ambiente, *Acordo de Paris 2015–2020*, Lisboa: Agência Portuguesa do Ambiente, pp. 3–6.

Steffen, W., Broadgate, W., Deutsch, L., Gaffney, O., & Ludwig, C. (2015). The trajectory of the Anthropocene: The great acceleration. *The Anthropocene Review*, *2*(1), 81–98.

Tavares, R. (2016). *Paradiplomacy: Cities and states as global players*. Oxford University Press.

The GEF (n.d.). *The global environment facility*. Available at https://www.thegef.org/ [Accessed on january 15, 2025].

The World Bank (2023). *O Banco Mundial em São Tomé e Príncipe*. Available at https://www.worldbank.org/pt/country/saotome/overview [Accessed on may 20, 2023].

Vila Nova, Carlos (2021). *Discurso do Presidente da República Democrática de São Tomé e Príncipe na Conferência das Partes das Nações Unidas sobre Alterações Climáticas*, Glasgow, December.

8

Navigating Education in Gaza

The Dual Impact of War and Climate Crises

Shadi Albanna

8.1 Introduction

Education is a key pillar in building societies and achieving sustainable development, as it not only empowers individuals and enhances their abilities to face various challenges, but also creates a more stable and prosperous environment that allows them wider opportunities for growth and development, which contributes to improving the quality of life and building a stable society. However, in many parts of the world, this essential sector is under serious threat due to political crises, armed conflicts, and worsening climate change.

In recent decades, the world has witnessed an alarming increase in the number of people affected by armed conflicts and forced displacement, including large-scale displacement of refugees, worsening health crises, and environmental disasters caused by climate change, which have hindered the educational process and deprived millions of children of their right to education. According to international reports, these crises have put 222 million school-age children at risk of serious obstacles to their education, pushing education systems to the brink of collapse, especially in areas affected by conflict and disasters (Education Cannot Wait, 2022).

Studies indicate that armed conflicts have a devastating impact on the educational process, not only through the physical damage to educational institutions but also through the psychological and social effects that hinder students' learning and academic development. In a UNESCO report (2011) entitled "The Hidden Crisis: Armed Conflict and Education," it was emphasized that armed conflicts have serious consequences for the goals of Education for All, which calls for protecting the right to education and strengthening educational services for children and youth affected by conflict (UNESCO, 2011).

Other international studies have shown that education in conflict zones suffers from a sharp decline in the quality of curricula, a lack of educational resources, and high school dropout rates due to psychological and economic pressures. In a report by Save the Children (2022), it was confirmed that children in conflict zones suffer from psychological disorders such as post-traumatic stress disorder (PTSD), which affects their ability to learn and engage in the educational process effectively.

Gaza is one of the most complex regions where these challenges are most evident. Education there faces a dual challenge resulting from repeated wars and climate change, which hinders the educational process and limits students' opportunities to obtain sustainable and quality education. There is no doubt that the educational process in Gaza has witnessed a noticeable decline in recent decades due to repeated wars and the siege imposed on the Strip for more than seventeen years. However, the last war that broke out on October 7, 2023, was the most destructive and influential on the educational process, as it led to the near-total destruction of the educational infrastructure. The Israeli occupation destroyed nearly 90 percent of educational institutions, including universities and schools, while the remaining schools were transformed into shelters for the displaced, causing a complete halt to the educational process for over a year, except for some individual efforts by teachers who resorted to teaching students in tents or through electronic initiatives via platforms such as Zoom.

According to reports, the war caused extensive damage to 1,661 educational institutions, including 927 schools, universities, kindergartens, and learning centers that were completely destroyed, while 734 educational facilities were partially damaged. Furthermore, the war also resulted in the loss of 12,800 student martyrs, in addition to 800 educational staff martyrs, depriving 785,000 students from continuing their education at all levels (Government Media Office, 2025). This added new psychological and social challenges for students and teachers.

The psychological effects of war on children in Gaza have reached unprecedented levels. UNICEF reports that almost all children in Gaza need psychosocial support, with the number of psychologically affected children rising to more than one million.

A recent study showed that 96 percent of Gaza's children feel that their death is imminent, while about half of them wish to die due to the deep psychological impact of the ongoing aggression (UN Geneva, 2024; The Guardian, 2024) These figures reflect the extent of the psychological crisis among students, making the resumption of the educational process more complicated, as children need a safe and psychologically supportive learning environment before they can return to learning normally.

In addition to the direct destruction of Gaza's educational infrastructure, the Israeli war has exacerbated climate crises, making educational challenges even more complex. A study by the *Guardian* revealed that the first few months of the war produced more greenhouse gases than twenty countries produced in an entire year. The emissions generated by Israeli attacks on Gaza over two months were equivalent to burning at least 150,000 metric ton of coal. According to the study, 99 percent of the estimated 281,000 metric ton of CO_2 were emitted in the first sixty days after October 7, as a result of Israeli aerial bombardment and ground invasion of Gaza (The Guardian, 2024).

In addition to the destruction of the education sector and the environmental impacts of the war, Gaza's infrastructure and basic services were severely damaged, further complicating the climate crises. 3,680 kilometers of electricity networks were destroyed, 2,105 power transformers were damaged, along with 538 generators and 16,266 solar energy projects, resulting in an almost complete paralysis of electricity supply and negatively impacting all vital sectors, including education (Government Media Office, 2025).

In the water sector, 335 kilometers of water distribution networks were damaged, 496 desalination plants, 111 central water tanks, and 120 water tankers were directly damaged, and 717 out of 1,698 water wells were out of service, exacerbating the drinking water crisis. As for sewage networks, 655 kilometers were damaged, in addition to 65 treatment plants, which contributed to the deterioration of environmental and health conditions in the Gaza Strip (Government Media Office, 2025).

In terms of road infrastructure, transport networks were severely damaged, with 3,916 square kilometers of roads and streets destroyed, making it more difficult for residents to move around and for educational and service institutions to be rebuilt. This extensive damage not only hinders the reconstruction of schools and universities but also makes it difficult to provide a suitable learning environment for students, exacerbating the education crisis in light of the repercussions of the war and increasing climate change (Government Media Office, 2025).

The widespread destruction of electricity, water, and sewage networks in Gaza has made it even more challenging to resume the educational process due to the lack of basic resources and vital services. This situation has not only disrupted education but has also contributed to increasing the frequency and severity of disasters associated with extreme weather events, hindering children's access to schools.

Additionally, many children are forced to spend long hours in water filling queues instead of attending classes or participating in educational initiatives that have been established to try to restore the educational process in Gaza. According to a report published by WAFA, children spend hours waiting for water, depriving them of valuable time needed for their education (WAFA, 2024).

The education sector in Gaza is facing unprecedented challenges due to the dual impact of repeated wars and climate crises, which have disrupted the educational process on multiple levels. In addition to the physical destruction of educational infrastructure, school and university students suffer from deep psychological and social crises that affect their ability to learn, threatening the future of upcoming generations and making it necessary to understand the dimensions of these challenges and develop strategies to address them.

Studies show that armed conflict not only disrupts education but also creates an environment of fear, anxiety, and trauma that limits students' ability to focus and achieve (UNICEF, 2024). Additionally, the loss of teachers and students due to war, coupled with a lack of educational resources and economic challenges, makes rebuilding the education system even more complex (Save the Children, 2022). Furthermore, climate changes such as water shortages and power outages exacerbate these issues, making it even harder to provide a stable learning environment.

In this context, this research aims to analyze the psychosocial impacts of the war and climate crises on students in Gaza, assess the extent to which the destruction of infrastructure affects learning opportunities, explore the role of teachers in supporting students and alleviating psychological pressures, as many children rely on teachers as their main source of psychosocial support, and finally, suggest strategies to enhance the resilience of the education system by developing innovative solutions that help ensure the continuity of education in areas affected by conflict and environmental disasters.

8.2 Research Questions

- How have war and climate crises affected students' mental health and social interaction?
- What challenges do students and teachers face in continuing the educational process under these circumstances?

- How did teachers adapt to challenges, and what effective methods were used to support students?
- What strategies can be implemented to ensure the sustainability and resilience of the education system in Gaza?

8.3 Study's Significance

This research is of great importance as it relates to one of the most pressing issues of our time, namely the dual impact of war and climate crises on the education sector in Gaza. Education is a fundamental human right and a critical factor in sustainable development and social stability (UNESCO, 2015). However, wars and environmental disasters significantly disrupt the educational process, resulting in long-term psychological, social, and economic damage to students, teachers, and society as a whole (Save the Children, 2022).

8.4 Understanding the Psychosocial Impacts on Students

Studies confirm that children exposed to armed conflict and environmental disasters suffer from severe psychological disorders such as anxiety, depression, and post-traumatic stress disorder (PTSD), which affects their ability to learn and engage in the educational process effectively (Betancourt et al., 2020). According to a UNICEF report (UNICEF, 2024), almost all children in Gaza need psychosocial support due to the ongoing violence, making it necessary to analyze how these factors affect the learning and mental health of students.

8.5 Analyzing the Impact of the Destruction of Educational Infrastructure

UN reports show that the destruction of schools and infrastructure disrupts education for many years, as students need a safe and stable environment to be able to learn effectively (Education Cannot Wait, 2022) Statistics indicate that the recent war in Gaza destroyed nearly 90 percent of educational institutions, depriving more than 785,000 students from continuing their education (Government Media Office, 2025) Therefore, analyzing the impact of this destruction on the educational process can contribute to developing strategies for rebuilding and sustaining the educational system.

8.6 Exploring the Role of Teachers in Supporting Students

In the absence of formal psychological support, many students rely on teachers as a primary source of emotional and social support, which calls for understanding their strategies in dealing with these challenges and how to improve their capabilities in this area (Winthrop & Kirk, 2008).

8.7 Recommendations to Strengthen the Resilience of the Education System

Studies indicate that educational systems in conflict zones need to develop resilient educational models, such as online education, flexible curricula, and in-school psychological support programs (Burde et al., 2017). Therefore, the results of this research can contribute to the development of educational policies that adapt to the difficult conditions in Gaza, helping to ensure the continuity of education for children suffering from trauma and forced displacement.

8.8 Contributing to the Global Debate on Education in Conflict Zones

Education in conflict zones is a central topic on the agendas of international organizations such as UNESCO, UNICEF, and the World Bank, as it is one of the key elements to achieve the United Nations Sustainable Development Goals (SDG 4) related to inclusive, equitable, and quality education for all (United Nations, 2015). By providing a comprehensive analytical study on Gaza, this research can contribute to the development of international policies that are more responsive to the needs of children in conflict zones.

8.9 Theoretical Framework

8.9.1 Overview of the Education System in Palestine

The education system in Palestine is one of the most important pillars that contribute to building national identity and strengthening cultural and political resistance, considering the exceptional circumstances imposed by the occupation and ongoing conflicts. For decades, education has been a means of establishing the values of unity and belonging, as it has constituted a field of cultural and political struggle in addition to its developmental role. In Palestine in general, and the Gaza Strip in particular, the system consists of several educational levels, including public schools, UNRWA schools that serve refugees, and higher education institutions that seek to qualify national cadres despite the restrictions imposed on resources and infrastructure.

The educational process in Palestine has been affected by many factors, including the historical and cultural heritage that motivates the preservation of national identity in the face of attempts to marginalize and occupy. The educational system has witnessed a series of reforms and initiatives aimed at integrating modern curricula with established educational traditions, which shows the desire of Palestinian society to adapt to global changes while maintaining its national characteristics. International institutions have played an important role in supporting education in the Palestinian territories, whether by providing technical and financial assistance or through psychological rehabilitation support programs for students and teachers who have experienced trauma (UNESCO, 2021).

Despite reform and development efforts, the education system in Gaza remains exposed to compelling circumstances that affect its continuity. Curricula and teaching methods are affected by repeated interruptions caused by wars, which hinder the continuity of the educational process. In addition, the system highlights challenges related to the lack of resources and deteriorating infrastructure, making it imperative for policymakers to search for innovative solutions based on developing the capabilities of the education system to adapt to urgent difficulties.

International experiences in conflict environments have confirmed that the role of education is not limited to the transfer of knowledge and skills but extends to supporting the psychological and social health of students and providing an educational environment that contributes to enhancing personal and societal capabilities to face multiple challenges. This has led to the adoption of educational curricula and programs aimed at promoting the concept of resilience and self-reliance, which is a step toward building a system capable of overcoming critical circumstances and continuing to perform its vital role (UNRWA, n.d.).

8.10 The Impact of Repeated Wars on Education in Gaza

The Gaza Strip is facing serious challenges as a result of repeated wars and armed conflicts, which directly and indirectly affect various aspects of the educational process. On the physical side, intensive Israeli bombing leads to the destruction of school facilities, loss of books and educational supplies, and severe damage to infrastructure that makes the continuity of education a daily challenge. Such frequent interruptions disrupt the educational process and affect students' academic achievement, as children are exposed to long periods of abrupt interruptions, which affect their understanding of the curriculum and lead to the accumulation of educational gaps (UNICEF, 2024).

Psychologically, the ongoing war environment exposes students and teachers to several psychological traumas. Violent events lead to increased rates of anxiety, depression, and post-traumatic stress disorder. This psychological impact is not limited to the time period of the war but extends to the post-war period, which requires continuous therapeutic intervention and psychological support to rehabilitate those affected and restore their educational capabilities. Many studies have indicated that the traumatic environment contributes to poor concentration and attention in students, which negatively affects the quality of education and increases school dropout rates (Aldabbour et al., 2024).

The impact of wars also highlights the erosion of trust between the educational community and the authorities responsible for ensuring the continuity of education, which is reflected in the levels of community participation in the educational process. Repeated wars weaken the socioeconomic structure of society, making it difficult to implement reform and innovation programs that would improve the quality of education and provide a safe and stable learning environment.

8.11 The Impact of Climate Crises on Education in Gaza

In addition to the impact of war, Gaza faces significant environmental challenges resulting from the worsening climate crisis. Climate change has exacerbated water shortages, deteriorating air quality, and rising temperatures, negatively impacting the health of students, teachers, and school infrastructure. Extreme weather conditions such as frequent heat and cold waves disrupt classes and cause serious damage to educational facilities, requiring ongoing maintenance and reconstruction to ensure a healthy learning environment (World Bank, 2023).

Climate crises also contribute to an increase in absenteeism among students, as some are forced to stay at home for safety during adverse weather conditions, resulting in the loss of valuable learning information and increasing the education gap. Environmental instability also affects the ability of schools to adopt modern educational technologies that rely on electricity and the internet, as the availability of these services is affected by power outages and the deterioration of public services during climate crises.

On the other hand, some studies indicate that environmental sustainability concepts should be integrated into school curricula to enhance students' awareness of climate issues and equip them with the knowledge needed to deal with future environmental challenges. This would contribute to building a generation capable of facing the effects of climate change and working to achieve sustainable development despite environmental constraints (World Bank, 2023).

8.12 The Overlap Between the Impact of War and Climate Crises

The overlap between the impact of war and climate crises is a complex phenomenon that requires a multidimensional understanding based on theories of resilience and complex systems. Under what is known as the "stressor duality" principle, the education system in Gaza is under intertwined pressures: repeated wars lead to infrastructure destruction and psychological impact, while climate crises exacerbate the fragility of the system by affecting natural resources and the learning environment.

From a theoretical point of view, the concept of resilience theory helps to highlight the ability of the education system to adapt to multiple shocks, as the educational community can adopt effective strategies to rebuild and recover after each crisis. In the same vein, the Complex Systems Theory model explains how political, economic, and environmental factors interact to form a complex educational reality that is difficult to separate into individual components. Understanding these interactions is key to developing integrative policies that ensure the continuity of education despite waves of simultaneous shocks (Masten, 2001).

The overlap between wars and climate crises highlights the need for an integrated strategic approach involving all actors, from local authorities to international institutions and civil society. Efforts should focus on developing emergency response mechanisms that enhance the resilience of the system while adopting educational programs that take into account the psychological and environmental needs of students and teachers. These holistic policies are an essential step toward achieving sustainability in the education sector in Gaza and ensuring that it remains a key tool for the advancement of Palestinian society despite the successive challenges.

8.13 Methodology

This research adopts the Qualitative Methodology, which allows for an in-depth understanding of social experiences and phenomena in their natural contexts (Creswell & Creswell, 2017) This method was chosen due to its exploratory nature that provides an in-depth analysis of the psychosocial factors that affect students in Gaza as a result of the war and climate crises. Qualitative research is characterized by its ability to extract meanings and patterns from non-numerical data, making it suitable for the study of psychosocial influences on the educational process (Patton, 2014; Denzin & Lincoln, 2011)

8.13.1 Data Collection Tool: Interviews

Semi-Structured Interviews were adopted as the main data collection tool, as this method allows flexibility in exploring teachers' views and experiences regarding the educational challenges faced by students in Gaza after the war and climate crises (Kvale, 2007). This tool provides the opportunity for participants to express their opinions freely, with questions directed in a way that helps reveal psychosocial influences and identify strategies used to support students.

8.13.2 Data Collection Details

1. **Research sample:** 12 teachers from different educational institutions in Gaza were selected using purposive sampling to ensure diversity of experiences and perspectives.
2. **Method of interviewing:** Interviews were conducted either in person or via virtual means (e.g., Zoom), depending on the security conditions and accessibility of the participants.
3. **Interview structure:** Interviews consisted of open-ended questions focusing on students' mental health, educational challenges, the impact of infrastructure destruction, and psychosocial support strategies.
4. **Data analysis:** The data was transcribed and analyzed using Thematic Analysis to extract recurring patterns and understand the factors influencing the educational process, using manual coding to ensure accuracy and consistency (Braun & Clarke, 2006)

8.14 Data Analysis and Discussion

This chapter aims to analyze the results of interviews with teachers about the impact of war and climate crises on the educational process. Data was collected from (twelve) male and female teachers from different schools and educational stages (see Table 1). The results were classified according to multiple themes, including the psychological and social impact on students, the challenges facing education, and possible methods of adaptation and support to ensure the sustainability of the educational process.

8.14.1 How Have War and Climate Crises Affected Students' Mental Health and Social Interaction?

Most teachers in this study confirmed that wars and climate crises have caused severe psychological disorders in students, as many of them suffer from disorders such as anxiety, depression, post-traumatic stress disorder, and chronic stress. These issues do not stop at the psychological aspect only, but extend to noticeable behavioral changes, such as aggression, social isolation, loss of concentration, and mental distraction, which reflect negatively on academic achievement and make it difficult for students to continue learning in an unstable environment.

Table 1 Sample Distribution of Study Participants

No	Teacher Name	Age	Gender	Years of Experience	Teaching Level	Specialization	School Type
1.	A	44	Female	1	Primary	Guidance & Psychological Health	Private
2.	B	35	Female	10	Primary	English	Private
3.	C	34	Female	1	Primary	Arabic	Private
4.	D	32	Female	10	Primary	Arabic	Private
5.	E	35	Female	10	Primary	Not Specified	Government
6.	F	58	Male	33	All Levels	Science & Math	Government
7.	G	50	Male	20	Secondary	Mathematics	Government
8.	H	39	Female	7	Primary	Special Education	Private
9.	I	42	Male	15	Secondary	Islamic Education	Government
10.	J	37	Female	15	Primary & Secondary	Science	Private
11.	K	40	Female	10	Primary	Social Studies	Private
12.	L	38	Female	14	Primary	General Education	Private

Most scientific research supports this view, as studies indicate that children exposed to such crises are more likely to suffer from profound psychological disorders. According to Betancourt et al. (2020), children who grow up in environments affected by armed conflict often suffer from post-traumatic stress disorder (PTSD), affecting their ability to adapt emotionally and behaviorally. Masten & Narayan (2012) explained that repeated exposure to trauma hinders the development of children's social and cognitive skills, leading to lower academic performance and increased behavioral issues such as aggression or social isolation, which was also observed by teachers in this study.

Among the most common behavioral changes in children affected by crises, aggression and social withdrawal stand out. Thabet et al. (2004) showed that children exposed to war trauma tend to show elevated levels of violence or isolation, often becoming more prone to anger and behavioral outbursts, or withdrawing completely from interacting with others.

Betancourt & Khan (2008) found that children who have experienced natural disasters or armed conflicts suffer from chronic feelings of fear and lose their sense of security, which leads to difficulty in building healthy social relationships within the school environment. These factors significantly hinder the learning process, as children become less able to focus and participate in classroom activities, which increases the challenges faced by teachers in providing an effective learning environment.

In addition to behavioral issues, one of the most common effects among students affected by crises is reduced ability to concentrate and comprehend. According to Shields & Cicchetti (2001), chronic stress resulting from trauma affects the executive functions of the brain, leading to impaired attention, shortened memory span, and difficulty absorbing new information. This is confirmed by UNESCO (2019), which indicated that children in conflict zones suffer from significant learning gaps due to the impact of trauma on their cognitive abilities. Therefore, it becomes necessary to develop supportive strategies within the school environment to help these students adapt to psychological pressures and improve their academic performance, ensuring that they have a better chance of continuing their education despite the challenges they face.

As for the behavioral changes in students, according to the teachers' responses, the most prominent behavioral changes in students were as follows:

- **Aggression:** Many teachers pointed to an increase in students' aggression, as they became more violent toward their classmates and tended to argue, shout, and fight repeatedly, even for minor reasons.

"Students have become very aggressive, attacking each other for the slightest reasons, and it has become difficult to control them inside the classroom."

Increased aggression among students exposed to wars and climate crises reflects the impact of trauma on their social behavior. Several studies indicate that children who face repeated trauma show increased aggressive responses, becoming more impulsive and prone to fights and arguments, even in simple situations. Trauma exposure at an early age affects children's emotion regulation, leading to increased violent behaviors as a result of their inability to express stress in healthy ways. In addition, aggression in children affected by crises may be an adaptive response to an unstable environment, as they adopt more aggressive behaviors to protect themselves or to cope with negative emotions.

- **Isolation and fear:** Some teachers report that a large proportion of students have become isolated and antisocial, especially those who have lost family members or experienced direct trauma.

"There are students who refuse to talk or participate in class after losing their families or homes."

Isolation and fear are among the most prominent behaviors exhibited by children who have experienced direct trauma, such as the loss of family members or forced displacement. Children who go through severe psychological experiences often tend to withdraw socially and prefer isolation as a result of increased feelings of anxiety and insecurity. Khamis (2019) showed that refugee or displaced children have higher rates of emotional disturbances, making them less able to interact with their peers or participate in classroom activities. This behavior is linked to a chronic fear of losing others or experiencing further trauma, making children more sensitive to new environments or changes in their social surroundings.

- **Difficulties with concentration and comprehension:** Teachers report that many students suffer from poor short-term memory, distractibility, and constant wandering, which has led to a significant drop in their academic level.

"Some students are unable to concentrate for more than 10 minutes in class, which makes learning very difficult."

Continued exposure to stress and trauma significantly affects the cognitive abilities of students. Children who experience chronic psychological stress have impaired short-term memory functions, leading to a reduced ability to concentrate and difficulty absorbing new information. Students in crisis-affected environments often suffer from sleep disorders, causing mental fatigue that negatively impacts their academic performance. In addition, sleep disorders, such as insomnia and trauma-induced nightmares, directly affect academic achievement, with children experiencing difficulty waking up in the morning, poor comprehension in class, and higher rates of absenteeism.

- **Sleep disturbances and bedwetting:** Some teachers have reported that many students experience frequent nightmares, difficulty sleeping, and bedwetting as a result of trauma.

Sleep disturbances and bedwetting are common symptoms among children who have experienced severe trauma. Children who go through traumatic experiences often have trouble regulating their sleep patterns due to chronic anxiety and stress, which is reflected in their academic performance and behavior in the classroom. Bedwetting is also a common psychological response in children affected by trauma, reflecting their lack of safety and control, especially in unstable environments. These issues not only affect the physical and mental health of students but may also lead to their reluctance to participate in social activities for fear of embarrassment, further isolating them within the school environment (Refugee Trauma Help, n.d.). Ongoing psychological support and early intervention programs are necessary to help these children overcome the effects of trauma and promote their psychological stability.

The results of the interviews showed that the impact of crises varies depending on the age group:

- Younger children (primary school): Show excessive fear, excessive attachment to teachers and parents, and sleep and speech disorders.
- Older students (middle and high school) tend to be violent, apathetic, depressed, or even consider dropping out of school due to a loss of educational motivation.

8.14.2 What Challenges Do Students and Teachers Face in Continuing the Educational Process Under These Circumstances?

The responses of teachers in Gaza illustrate the significant challenges facing the educational process under the exceptional circumstances in Gaza. Teachers explained that schools have suffered direct and indirect damage that led to the destruction of infrastructure, with some schools turned into shelters, which hindered the effective resumption of the educational process.

They also pointed out that the destruction of roads and transportation has made it difficult for students and teachers to reach schools, which exacerbates the suffering of the educational process. One teacher stressed: "The biggest challenge is the lack of schools. What is available are tents that are not equipped to accommodate students." In addition, teachers pointed to a severe shortage of educational resources; such as textbooks, educational tools, and basic supplies like graves and blackboards, in addition to the power and internet outages that has made distance education ineffective, which creates a knowledge gap that negatively affects students' achievement.

On the other hand, Gaza teachers indicated that the financial challenges are accompanied by severe psychological pressures that affect their performance and ability to provide psychological support to students. They live in the same environment in which students are exposed to trauma, which makes them vulnerable to problems of anxiety, stress, and emotional exhaustion. In this context, one teacher stated: "We try to do our best, but we suffer from the same psychological conditions as our students, teachers need psychological support too." Teachers stressed that low salaries and high costs of living increase these pressures, which necessitates the provision of continuous psychological and institutional support to improve working conditions and ensure the continuity of high-quality education in light of the current challenges.

The United Nations Office of the Special Adviser on Africa (2024) study confirms that the lack of textbooks and basic supplies, such as desks and blackboards, limits teachers' ability to deliver lessons effectively, leading to a decline in students' academic achievement. The lack of educational tools also creates a knowledge gap among students, as children in poor areas face greater difficulties in accessing educational content than their peers in more stable environments. These obstacles reinforce inequalities in learning opportunities and lead to lower enrollment and retention rates.

On the other hand, the lack of digital infrastructure and power and internet outages are major factors that undermine the effectiveness of distance education, especially during crises. Accordingly, the lack of electronic devices or stable internet connections deprives many students of access to digital educational resources, which leads to a decline in learning levels and widens the educational gap between different social groups, reducing their chances of academic success in the long term. Therefore, providing the appropriate infrastructure and modern technology is essential to ensure the continuity of the educational process and promote equal opportunities for all students.

In addition to these financial challenges, teachers face severe psychological pressures due to being in the same environment as students, as they are forced to deal with problems and traumas on a daily basis that negatively affect their mental health. Teachers' constant exposure to students' suffering and repressed feelings leads to an accumulation of feelings of anxiety and stress, which increases the likelihood of emotional exhaustion and psychological fatigue. This situation makes it difficult for them to maintain their psychological stability, which is directly reflected in the quality of the educational process and their ability to provide the required support to students.

These intertwined challenges, both material and psychological, require a comprehensive and sustainable response from stakeholders. It is essential to provide financial and psychological support to teachers, improve school infrastructure, and provide the necessary educational and technical resources. Finding effective solutions to these problems not only ensures the continuity of education but also contributes to reducing the educational gap and enhancing students' chances of success in a more stable and supportive educational environment.

8.14.3 How Did Teachers Adapt to Challenges, and What Effective Methods Were Used to Support Students?

Teachers' responses show that there are significant efforts to enhance students' ability to cope with the psychological pressures resulting from the difficult circumstances imposed by conflict and crises. Teachers explained that, despite the great challenges, they sought to invest in various strategies aimed at supporting students psychologically and mitigating the effects of trauma, which helps in the continuity of the educational process even under exceptional circumstances.

In terms of strategies to support students psychologically, teachers indicated that they relied on innovative teaching methods that contribute to expressing feelings and relieving stress. Some of them used learning methods through play, such as drawing, coloring, interactive drama, and storytelling, to provide an environment that allows students to express their feelings freely. Relaxation techniques, such as deep breathing exercises and meditation, were also applied, which helped reduce levels of anxiety and stress. In this context, one teacher said: "I tried to use play and recreational activities to help students overcome their negative feelings, and some of them gradually began to return to interaction." These strategies demonstrate teachers' keenness to create a safe and stable classroom environment that contributes to enhancing emotional communication between students and teachers. This is supported by Ginsburg's (2007) study, which showed that play is of great importance in improving students' psychological health and development.

As for improving the learning environment, Gaza teachers indicated that there are concrete field initiatives aimed at providing educational alternatives in light of the deteriorating infrastructure. They have set up educational tents in the most affected areas to provide a temporary and safe environment for the continuation of the educational process, and they have sought to integrate interactive activities and modern methods to compensate for the shortage of traditional books and educational supplies. One teacher expressed the importance of these solutions, saying: "We must find innovative solutions to ensure the continuity of education despite all the challenges." These initiatives reflect the spirit of creativity and challenge among teachers, who are striving hard to ensure the provision of sustainable education that addresses the needs of students under the current circumstances.

Teachers also stressed the importance of cooperation with the local community as one of the main pillars of supporting the educational process. Charitable societies and local organizations contributed to providing direct financial support to students through the distribution of school bags and educational books, which helped ease the burden on families. Workshops and training courses were also organized for teachers to enhance their skills in dealing with students affected by psychological trauma, which enabled them to acquire the necessary tools and strategies to support students on the psychological and educational levels. This close cooperation with the local community confirms that solidarity and mutual support play a crucial role in enhancing the continuity of the educational process and achieving a more stable and cohesive classroom environment, even in the most difficult circumstances.

8.15 What Strategies Can Be Implemented to Ensure the Sustainability and Resilience of the Education System in Gaza?

Based on teachers' responses, they explained that maintaining the sustainability and resilience of the education system in times of crises requires adopting innovative strategies that are appropriate to the exceptional circumstances they are experiencing. Teachers stressed the importance of enhancing digital education by providing free e-learning platforms that ensure students' access to lessons even in emergency situations. They also pointed to the need to distribute tablets and e-books to compensate for the lack of traditional educational resources, which contributes to overcoming the educational gap resulting from the challenges imposed by the conflict and the deterioration of infrastructure.

Furthermore, teachers stressed the need to build more crisis-resistant schools by establishing mobile schools or temporary educational camps to accommodate students in the most affected areas and to accelerate the reconstruction of damaged schools as soon as possible. They also stressed the importance of enhancing psychosocial support for students by appointing psychological counselors in schools to provide ongoing support and integrating psychosocial support programs into the curriculum to ensure a sustainable positive impact. As one teacher said, "Students need sustainable psychological support, not just temporary initiatives that will disappear after the crisis." These strategies demonstrate the commitment of Gaza's teachers to overcome challenges and achieve continuity of education despite difficult circumstances.

8.16 Study Recommendations

To address recurring crises, it is necessary to enhance psychosocial support for students by integrating psychosocial support programs within schools, providing an environment that helps them deal with traumas such as anxiety and depression. In addition, training teachers on trauma-sensitive education enables them to understand students' psychological needs and deal with their behaviors flexibly. Furthermore, psychological counseling services should be provided within schools by appointing specialized counselors to support students psychologically and emotionally, as well as encouraging the creation of support groups led by students themselves, which helps them express their feelings and feel a sense of belonging. Incorporating relaxation and stress management techniques, such as meditation and deep breathing, is also effective way to help students deal with daily stress and enhance their academic focus.

Regarding the challenges related to infrastructure and resources, it is necessary to rehabilitate damaged schools and equip them with basic facilities such as classrooms and desks, thus providing an appropriate learning environment. Access to educational resources should also be improved by providing textbooks and digital supplies to students and teachers to ensure continuity of learning. In addition, providing safe transportation for students and teachers is essential to ensure their regular access to schools, especially in unstable situations. Psychological and professional support for teachers is also essential to improve their performance and help them interact effectively with students. To compensate for any disruption in education, distance learning capabilities should be enhanced by developing digital educational platforms that allow for the continuation of study in emergency situations.

To help teachers and students cope with difficult situations, it is essential to encourage alternative teaching methods such as interactive learning, storytelling, and drama to make education more engaging and accessible. Temporary learning centers can also be set up in affected areas to ensure that education continues during crises. Enhancing school-community collaboration is essential to create a stable learning environment. In addition, stress-relieving activities through sports and play should be implemented to help students relieve stress. To improve students' performance and concentration, it is important to provide school feeding programs that ensure students receive nutritious meals that support their physical and mental health.

To ensure sustainability and resilience of the education system, flexible education policies should be developed that can quickly adapt to crises, reducing the impact of conflicts and natural disasters on the educational process. Investing in digital infrastructure and providing students with the internet and tablets contributes to facilitating access to education. It is also important to build schools that can withstand crises and disasters to ensure that education continues in all circumstances. In addition, programs that enhance psychological resilience can be integrated into the curriculum, helping students adapt to different challenges. Finally, cooperation with local and international organizations is essential to secure the funding and resources needed to support education in Gaza and ensure its continuity under harsh conditions.

To ensure the development of the education system in Gaza and to meet the increasing challenges, there is an urgent need to conduct more studies on the impact of recurring crises on students' academic achievement and mental health. Researchers could focus on analyzing the psychosocial coping strategies used by students and teachers in crisis settings, and the effectiveness of current psychological support programs in mitigating the effects of trauma. Innovative educational approaches that are suitable for unstable conditions could also be studied, such as technology-based learning or interactive teaching methods that do not rely on traditional resources.

In addition, it is important for researchers to explore the sustainability and effectiveness of distance learning platforms in areas affected by conflict and natural disasters, and to analyze the digital gaps that hinder equitable access to e-learning. Studies could also address the future of educational infrastructure, in terms of designing more resilient and crisis-resilient schools. Furthermore, research can be conducted on the role of partnerships between educational institutions and international organizations in supporting and funding education under difficult conditions, which helps to develop long-term strategies to ensure the continuity of the educational process and achieve the maximum benefit for students and teachers.

8.17 Conclusion

The findings of this research clearly indicate that wars and climate crises present severe challenges to the educational process in Gaza. They adversely affect the psychological and social well-being of students, leading to decreased concentration and academic performance, while also triggering behavioral changes such as aggression and isolation. Teachers, who bear the dual burden of teaching and providing psychosocial support in a traumatic environment, face considerable challenges exacerbated by the destruction of educational infrastructure and a shortage of resources.

To overcome these challenges, the study recommends:

- Strengthening psychosocial support programs within schools.
- Training teachers in trauma-sensitive teaching methods.
- Providing flexible learning environments through mobile schools and digital education.
- Enhancing collaboration between educational institutions, local communities, and international organizations to secure the necessary resources.

Ultimately, a holistic and sustainable approach is required to ensure the continuity of education and to improve students' prospects for a stable and prosperous future, despite the harsh realities posed by ongoing wars and climate crises.

References

Aldabbour, B., Abuabada, A., Lahlouh, A., Halimy, M., Elamassie, S., Sammour, A. A. K., . . . & Nadarajah, S. (2024). Psychological impacts of the Gaza war on Palestinian young adults: A cross-sectional study of depression, anxiety, stress, and PTSD symptoms. *BMC Psychology, 12*(1), 696.

Betancourt, T. S., & Khan, K. T. (2008). The mental health of children affected by armed conflict: Protective processes and pathways to resilience. *International Review of Psychiatry, 20*(3), 317–328.

Betancourt, T. S., Keegan, K., Farrar, J., & Brennan, R. T. (2020). The intergenerational impact of war on mental health and psychosocial wellbeing: Lessons from the longitudinal study of war-affected youth in Sierra Leone. *Conflict and Health, 14*, 1–8.

Braun, V., & Clarke, V. (2006). Using thematic analysis in psychology. *Qualitative Research in Psychology, 3*(2), 77–101. https://doi.org/10.1191/1478088706qp063oa

Burde, D., Kapit, A., Wahl, R. L., Guven, O., & Skarpeteig, M. I. (2017). Education in emergencies: A review of theory and research. *Review of Educational Research, 87*(3), 619–658.

Creswell, J. W., & Creswell, J. D. (2017). *Research design: Qualitative, quantitative, and mixed methods approaches.* Sage publications.

Denzin, N. K., & Lincoln, Y. S. (Eds.). (2011). *The Sage handbook of qualitative research.* Sage.

Education Cannot Wait. (2022). *Global estimates: Number of crisis-affected children and adolescents in need of education support.* https://www.educationcannotwait.org/sites/default/files/2022-06/ecw_globalestimatesstudy_june2022.pdf

Ginsburg, K. R. (2007). The importance of play in promoting healthy child development and maintaining strong parent-child bonds. *Pediatrics, 119*(1), 182–191. https://doi.org/10.1542/peds.2006-2697

Government Media Office. (2025, February 2). *Press Release No. (749) issued by the Government Media Office.* [Telegram post]. . https://t.me/mediagovps/3474

Khamis, V. (2019). Posttraumatic stress disorder and emotion dysregulation among Syrian refugee children and adolescents resettled in Lebanon and Jordan. *Child Abuse & Neglect, 89*, 29–39.

Kvale, S. (2007). *Learning the Craft of Interviewing.* Sage publications.

Masten, A. S. (2001). Ordinary magic: Resilience processes in development. *American Psychologist, 56*(3), 227–238. https://doi.org/10.1037/0003-066X.56.3.227

Masten, A. S., & Narayan, A. J. (2012). Child development in the context of disaster, war, and terrorism: Pathways of risk and resilience. *Annual Review of Psychology, 63*(1), 227–257. https://doi.org/10.1146/annurev-psych-120710–100356

Patton, M. Q. (2014). *Qualitative research & evaluation methods: Integrating theory and practice* (4th ed.). Sage Publications.

Refugee Trauma Help. (n.d.). *What is trauma? How does our body respond to it?* Retrieved from https://www.refugee-trauma.help/ar/about-trauma/

Save the Children. (2022). *Global environmental sustainability report 2022.* Save the Children. https://resourcecentre.savethechildren.net/document/save-the-children-global-environmental-sustainability-report-2022

Shields, A., & Cicchetti, D. (2001). Parental maltreatment and emotion dysregulation as risk factors for bullying and victimization in middle childhood. *Journal of Clinical Child Psychology, 30*(3), 349–363. https://doi.org/10.1207/S15374424JCCP3003_7

Thabet, A. A. M., Abed, Y., & Vostanis, P. (2004). Comorbidity of PTSD and depression among refugee children during war conflict. *Journal of Child Psychology and Psychiatry, 45*(3), 533–542. https://doi.org/10.1111/j.1469-7610.2004.00243.x

The Guardian. (2024, December 11). *Death feels imminent for 96% of children in Gaza, study finds.* The Guardian. https://www.theguardian.com/world/2024/dec/11/death-feels-imminent-for-96-of-children-in-gaza-study-finds

UNESCO. (2011). *The hidden crisis: Armed conflict and education; EFA global monitoring report.* UNESCO. https://unesdoc.unesco.org/ark:/48223/pf0000190743

UNESCO. (2015). *Education 2030: Incheon Declaration and Framework for Action for the Implementation of Sustainable Development Goal 4.* https://unesdoc.unesco.org

UNESCO. (2019). *Global education monitoring report, 2019: Migration, displacement and education: Building bridges, not walls.* https://doi.org/10.54676/XDZD4287

UNESCO. (2021). *Global Education Monitoring Report 2021: Education for All in a Post-COVID World.* UNESCO Publishing.

UNICEF. (2024). *Children in Gaza: A generation under siege.* https://www.un.org/unispal/document/palestinian-children-under-siege-unicef-supporting-a-generation-of-children-longing-for-peace-report/

United Nations Geneva. (2024, February). *Children of Gaza: Thousands lost their parents, and a million need psychological and social support.* United Nations Geneva. https://www.ungeneva.org/ar/news-media/news/2024/02/90019/atfal-ghzt-alalaf-fqdwa-dhwyhm-wmlywn-bhajt-aly-dm-nfsy-wajtmay

United Nations Office of the Special Adviser on Africa. (2024). *Africa's development dynamics 2024: Policy brief 3.* United Nations. https://www.un.org/osaa/sites/www.un.org.osaa/files/files/documents/2024/publications/ads2024_policybrief3.pdf

United Nations. (2015). *Transforming our world: The 2030 agenda for sustainable development.* https://sustainabledevelopment.un.org/content/documents/21252030%20Agenda%20for%20Sustainable%20Development%20web.pdf

UNRWA. (n.d.). *Education in the Gaza Strip.* Retrieved February 22, 2025, from https://www.unrwa.org/activity/education-gaza-strip

WAFA. (2024, February 9). *Children of Gaza... in water queues instead of classrooms.* WAFA News Agency. https://www.wafa.ps/Pages/Details/103246

Winthrop, R., & Kirk, J. (2008). Learning for a bright future: Schooling, armed conflict, and children's well-being. *Comparative Education Review*, *52*(4), 639–661.

World Bank. (2023). *West Bank and Gaza: Country climate and development report.* https://documents1.worldbank.org/curated/en/099112823094030933/pdf/P179452029863f0e80a902031b18e72860c.pdf

9

Radio Journey

The Future of the Climate Program

Neşe Yenice

9.1 The Story of Our Program

I am the producer of the program "The Future of the Climate," which is broadcast live every Thursday from 16:05 to 16:30 on TRT Radio 1. Our story began in 2020. Why did I choose the topic of climate? Because the future of the climate is the future of all of us—of all people and all living beings in nature. Of course, I could have planned or proposed more entertaining and widely listened-to programs. However, my goal was not to remain indifferent to the global destruction that threatens both our world and our country. It was to take a small step for our future, to do something with the responsibility of a broadcaster, and to listen to science.

Between 2006 and 2012, I also produced a program titled Energy in Focus. The depletion of fossil fuels and the damage they cause to our planet had caught my attention. Starting from those years, my interest in renewable energy began to grow. When I proposed this program for 2020, it marked the start of another learning process for me. Climate change is one of the greatest challenges of our time. For this reason, through the program I started in 2020, I made it my mission to raise public awareness.

In The Future of the Climate, our goal, particularly as a public broadcaster, is to inform and raise awareness on this issue. In addition to enhancing environmental awareness and sharing accurate, objective, and systematic scientific data, we strive to make individuals and society more conscious through social responsibility projects.

We are all just tiny drops on this planet. Very small. Yet, at the same time, we are immense. Every person, in their seemingly tiny existence, can achieve great things. The program focuses on conveying to listeners the global and local impacts of climate change, ways to cope with these effects, and sustainable living practices.

9.2 What Have We Done?

In 2025, our program said "hello" to its sixth year. So, what have we accomplished during this time?

First, I became a student. To capture the audience's attention and convey information accurately, it's essential to read, research, and present everything within a certain framework, reflecting its relevance to society and individuals. I emphasized the importance of explaining to listeners that phenomena such as melting glaciers, vegetation regrowth, or two days' worth of rainfall in the Sahara Desert equaling what used to fall in fifty years are directly connected to Turkey and that what seems like a distant danger is actually close to home. Of course, the issues we face in our country have also been part of our discussions. We tried to highlight that all of these are pieces of a larger whole because climate change creates significant societal, environmental, and economic impacts on both global and local scales.

For this reason, we thoroughly researched every climate-related topic and invited the most qualified experts to our program. We hosted hundreds of guests—academics, representatives from non-governmental organizations, and climate activists—who shared their insights with us, whether in the studio or over the phone. We learned something new from each of them, and with every guest, our pool of knowledge grew.

We discussed water, air, soil, agriculture, food security, food safety, drought, climate migration, biodiversity, birds, trees, forests, plastics, carbon footprints, digital footprints, the circular economy, organic farming, waterless farming, ecological farming, lakes, seas, oceans, fisheries, climate modeling, climate adaptation, climate-resilient cities, rainwater harvesting, renewable energy, the sun, wind, biofuels, next-generation aircraft fuels, and recycling.

Our programs were enriched by the efforts and contributions of climate activists. We found hope in children and young people who reminded us, "There is no Planet B" and "We have no other planet to live on." We always followed the agenda and current developments. Important global meetings and initiatives, such as IPCC reports and their implications or COP summits, were also among our topics. We introduced books that discuss climate change and emphasize coexisting with nature. Surveys conducted on climate issues were also part of our discussions.

9.3 Awareness

While planning for tomorrow, next week, or the coming year, we must also plan for the future of our planet. I believe that our contributions, no matter how small, are incredibly important and valuable. Instead of saying, "What difference can my small effort make?" we can all become climate-friendly, climate guardians, climate activists, and friends of the Earth in our homes and workplaces.

I believe that awareness developed on an individual level can also have a significant impact on a macro scale. Small steps can turn into a massive avalanche, creating strong demands for decision-makers and prompting them to take action. Being a climate activist doesn't necessarily mean participating in street demonstrations. It involves planning and managing our lives accordingly and making this a part of our daily routine. We can start by turning off running taps, switching off unnecessary lights, separating waste for recycling, watering plants, using the blank side of paper, avoiding unnecessary printing, turning off computers and speaker systems when not in use, avoiding unnecessary clothing purchases, using existing ones to their full lifespan, repairing them, not over-shopping for food, buying only as needed, composting, growing plants at home or work, planting trees, eliminating single-use plastics from our lives, taking our own bags to the market, not running washing machines or dishwashers unless they're full, choosing lower temperature settings, avoiding littering, and not discarding waste on the streets. There's so much we can do.

I believe that we must start with ourselves to adapt to the climate and ensure that our words align with our actions.

In solving this issue, the media must also "adapt to the climate" and embrace a journalism-broadcasting approach that is "climate-conscious." In our country, there is a lot of productive output from the online media on this subject. We now see more climate-related news in the print and digital media, but it's still not enough. Climate change is often only discussed when extreme weather events like hurricanes, tornadoes, and floods occur, and then it's quickly forgotten, as if it's a finished matter. However, this is an ongoing issue that is part of our daily lives, and its effects will continue to grow. The press must keep this issue in the foreground and should never tire of discussing it. The more it is talked about and repeated, the more people will learn and become knowledgeable and willing to make demands of decision-makers. When communicating climate issues to the public, it's important not to create fear and panic but to approach it in a way that conveys the seriousness of the situation and offers solutions.

9.4 Conclusion

The nature-friendly efforts I just mentioned are actions that are synonymous with not throwing our lives away, not wasting our world. Starting with ourselves, raising our children as climate-conscious individuals who are at peace with the world must be our most important responsibility.

Every week, we close our program with the following words:

"It's the perfect time to take action to stop climate change and secure the future of our planet. By joining hands, living in harmony with nature, and listening to science, we can overcome this issue. For a better, cleaner, more livable world, we are with you for the climate every Thursday."

Yes, I believe that by listening to science, hearing the voice of the world, and joining hands, this problem can be stopped. It's not the time for hopelessness, but for action. It's not the time to stop, but the time to run. It's the time for curiosity and research. It's the time to be the most nature-friendly consumer we can be.

In parallel with the negative trends indicated in the IPCC reports, it's time to take faster, more widespread, and more intense steps. When we understand that the universe does not belong solely to humans, but is a link in the biological diversity shared with all living beings, most of the problem will be solved.

In these days when we talk a lot about the concept of sustainability, I believe that it can be achieved through "sustainable curiosity" on an individual level.

At the beginning of every program, we cover important climate news from the past week. Unfortunately, there are often troubling reports from around the world: floods, fires, deforestation, the drying up of rivers, problems caused by extreme warming of oceans and seas, and the melting of glaciers . . .

I hope that after 2030, climate news will change. I imagine that the climate programs aired in those years will be filled with topics that reflect the beauty of our world, showcasing the right practices and providing uplifting perspectives. I hope more people will become sensitive to these issues, and that we can move toward a collective awareness in the fight against climate change. I wish for encouraging developments such as the increased use of renewable energy sources, the revival of ecosystems, and the protection of nature to be at the forefront of our agenda.

I believe that no step is ever in vain, and that every step can create societal transformation. Of course, there are times when I feel hopeless. During those times, I say,

"Keep pushing forward with the shovels!"

And I imagine the climate news after 2030 . . .

"The sun is shining across the world . . . The transition to renewable energy has accelerated. In the near future, 100% renewable energy will be possible.

Single-use plastics have been removed from our lives.

The world is lush and green. Our forests are increasing. The seas and oceans are deep blue.

Island nations have been saved from rising waters. Climate finance is flowing to underdeveloped countries from all over the world.

Climate activists are very happy. They have set off on a world tour by bike.

The melting of glaciers has stopped. The temperatures in the oceans have returned to normal.

The seas are full of fish. Our rivers are flowing stronger.

We managed to stay within 1.5 degrees."

I think of the world as a polyphonic choir. Just like how soprano, alto, tenor, and bass voices in a polyphonic choir reach a single sound with different melodies, the world should unite with different voices to form a single voice. In the choir, we reach "truth, good, and beauty" by listening to each other and singing together.

The fight against climate change should be the same—solving problems by listening to each other and talking together. For a more livable world, let's all work together, globally, "Keep pushing forward with the shovels!"

Part III

Governance and Leadership in Climate Diplomacy

Quo Vadis

10

Bringing the Functional Aspect of the Earth to International Law and Institutions

Paulo Magalhães

10.1 Introduction

To date, the strategy to combat climate change has had two main distinct objectives. While Article 2 of the 1992 Rio Convention,[1] focused its axis of action on the total concentrations of greenhouse gases (GHG) in the atmosphere, that is, the total ***stock*** **of** CO_2 accumulated in the Climate System, the Paris Agreement's goal[2] is not to exceed "a global ceiling on ***GHG emissions,***" *that is,* a strategy focused on ***flows.*** A strategy predominantly focused on mitigating current emissions—***flows***—*does* not allow societies *to start a process to* curb the excessive CO_2 accumulated in the atmosphere—***stock***—to keep the 1.5 degrees Celsius Paris goal achievable,[3] or to avoid irreversible "tipping points in which the feedback process becomes self-perpetuating after a critical threshold is crossed (. . .)The Hothouse Earth."[4]

[5]Changing the current trajectory and avoiding irreversible tipping points,[6] depends on large-scale human proactive positive actions that are able to produce "*negative feedbacks that could maintain the Earth System in Holocene-like conditions— notably, carbon uptake by land and ocean systems.*"[7]

This means that it is absolutely necessary for this negative feedback to be carried out on a large scale even before neutrality. As UNEP's emissions gap report explains, "all these scenarios would cause debilitating impacts to people, planet and economies, and necessitate future costly and large-scale removal of carbon dioxide from the atmosphere to bring down the overshoot of the Paris targets."[8] In other words, we are faced with a problem of a stock of excess CO_2

accumulated in the atmosphere; merely reducing emissions and/or addressing flows is insufficient to avoid tipping points.

Climate change is frequently regarded as a Tragedy of Commons[9] on a global scale,[10] or a Social Dilemma, where cooperation could make everyone better off, but incentives induce individuals, businesses, and nations to keep on emitting GHG.[11] How is it possible to avoid this dilemma and produce "biogeophysical feedbacks within the system that can maintain it in a given state (negative feedbacks)," and shift the current hothouse trajectory if "currently there are no economic mechanisms designed to pay for negative emissions"[12] or global ecosystem services restoration?

Nowadays, removing the excess accumulated *stock* of CO_2 in the atmosphere, without a direct link to offsetting emissions, corresponds to a removal carried out *in the qualitative/functional aspect* of the planet, which is devoid of any legal value. Thus, the benefits produced in the functioning of the Earth System socially disappear in a legal void, where no rights are generated for those who produce this benefit.

We face the dual challenge of internalizing negative externalities or costs (overexploitation), but also internalizing positive externalities or benefits (negative emissions, ecosystem services), which leads to a "*collective action dilemma*." According to Olson[13] "collective action is a problem because the costs of contributions are concentrated while the benefits are diffuse." The fundamental problem is the same in both situations: all individuals continue to have incentives to increase the use of the common good because the private benefits are greater than the share of the public costs that may fall to them ("private costs"); on the other hand, the *positive inputs* generate individual costs that are not compensated and can only be used in a very small part by those who took the initiative to create the collective benefit. In this logic, no one has a rational incentive to generate benefits, and everyone tries to exploit the good as quickly as possible before the rest of the group does, even if the results ("public costs") are disastrous for the community as a whole. This picture leads Elinor Ostrom to state that Olson has explained not a dilemma, but "a ***theory of*** **collective** inaction."[14]

For Hardin, this dilemma and the resulting tragedy primarily stem from "(…) insufficient delimitation of property rights that translates into overexploitation of natural resources.[15]

Today, from a legal point of view, the planet is just a territory with a total surface area of 510 million square kilometers divided between states, in which the *Global Commons* are just the leftover territories from these divisions.[16] This unidimensional approach excludes the qualitative/functional aspect of the planet from the legal object, and this invisibility has transformed the two major circulation systems, "the oceans and the atmosphere, into the garbage dumps of the industrial age."[17] The biochemical quality of seawater and the air is decisive in the biogeophysical processes that condition and determine the operational mode of the Earth System—***the functional aspect.*** The fact that the current legal approach is exclusively territorial, it mixes in the same concept of territory measured only in square kilometers, the quality/functional aspect that is measured by a set of quality indicators.

The outcome is a process that legally conceals the functional aspect, resulting in an undefined legal gray area. Taking into account that the territory is divisible and can be subject to national jurisdictions, and the functional aspect is global, indivisible, and cannot be subject to national jurisdictions, this results in the legal invisibility of the functional aspect, which is treated as "external" to the international legal system, and consequently, external to the economy—"the intangible place," where the economy sends "positive externalities," such as ecosystem services, and "negative externalities," like pollution—the "externalities dump."

The current inability to represent the factual overlap between the planet's functional aspect (biophysical systems on Earth) and its territory has also prevented us from reaching a consensus on the definition of goods that we usually call "global commons": "*There is much talk and activity about 'global commons,' but there is exceedingly little consensus on its definition. Different disciplines approach the discussion with widely varying assumptions, axioms, and goals, producing a body of writing that is inconsistent, if not incoherent.* (. . .) Global Commons: Is a definition possible"?[18] Because this "issue itself touches on some deep human concern"[19] and "we urgently need a new global approach to safeguard critical Earth System regulating functions more effectively and comprehensively,"[20] it is critical for the future of humanity "governing biophysical systems on Earth upon which the world collectively depends."[21] The fact that the functional aspect of the planet being "external" to the legal system has turned it into nobody's stuff, incapable of generating rights and duties, and of building a system that organizes its use and ensures its maintenance makes it impossible to explain functions and dynamics within a static territorial object. When what is at stake is the life support system itself, the outcome is an "existential threat to humanity."[22] This represents a genuine theoretical gap: "a *gap between theorization of reality and the actual reality of the ecological dynamic of the biosphere,*"[23] which inevitably results in a huge haze: "*Legally described as non-state space, outside territorial sovereignty, the global commons are not protected as domains that belong to everyone,* res communis. *Instead,* international law treats them as domains that belong to no one, res nullius *(. . .).*"[24] The 'state of nature' for the global commons is *res* nullius."[25]

If we refer to the atmospheric and oceanic circulation observed during the Holocene period that supported human civilizations, we can identify a certain pattern of functioning, a relatively stable "software,"[26] making it possible to approach the functional aspect independently of territorial considerations, "the hardware" and using this analogy, initiate a process to explain the gap. In this line of thought, in 2021, the United Nations International Law Commission (ILC)[27] clearly identified the problem: "The atmosphere and airspace are two different concepts that must be distinguished. (. . .) The atmosphere, as an 'envelope of gases' that surrounds the Earth, is dynamic and fluctuating, with gases that constantly move without regard to the territorial boundaries of States. The atmosphere is invisible, intangible and indivisible." Airspace refers to "a static and spatial institution over which the State, within its territory, has complete and exclusive sovereignty." But as "*atmospheric movement has dynamic and fluctuating characteristics (. . .)" it is considered to be the "functional aspect (. . .) that involves the movement of air on a large scale.*" By analogous reasoning, a similar distinction should be applied between the biochemical composition of seawater and the global circulation of the oceans and the maritime spaces under the sovereignty or jurisdiction of States.

Combining the need to proceed to this distinction put forward by ILC with the previous analogy, we explore the reasons why we intuitively refer as *global commons—res communis*—to vital goods that we all depend upon and share, yet they are treated as belonging to no one—*res nullius*. This step is critical in order to be able to decipher and explain this structural obstacle of uncertainty.

For rights and obligations to exist, there needs to be a subject (holder) and an object on which these rights and duties can be applied. "**How can a good that belongs to no one be subject to a legal regime?**."[28] Today's territorial spaces of the oceans and the atmosphere already possess well-defined legal statuses (territorial spaces subject to state jurisdiction and outside state jurisdiction), but these territorial statuses conceal the fact that they simultaneously constitute the territorial space in which these two major circulation systems operate. In this case, we have a clear overlap between the biochemical quality of water and air (which are global and indivisible) and the legal status of territorial spaces (which are local and divisible) in which they circulate. Therefore, the garbage dumps of humanity are the biochemical quality of water and air of these major circulation systems, regardless of the legal status of the territories through which they circulate. This overlap of two different realities—one qualitative/functional and another static and territorial—is not legally recognized and therefore is incapable of generating legal effects. The current exclusively territorial approach hides and makes legally invisible the global functional aspect of the planet.

Recognizing the functional aspect autonomously in relation to the territory requires defining its character, identifying the elements that describe it, delimiting it, and quantifying it with precision to enable its conceptual representation. However, this legal evolution is critical, as from the perspective of human societies perspective, it is not possible to boost an human-created "pathway of Earth System stewardship guided by human-created feedbacks to a quasi-stable human-maintained basin of attraction,"[29] if these positive impacts occur on assets lacking legal recognition, and therefore are incapable of producing precise legal effects. Currently, the efforts to generate positive impacts disappear into a legal void, meaning that they become economically invisible, and making their production a purely altruistic act, which will never be on the scale of what is needed.

As in any human enterprise, the most basic requirement must be in place: a proper legal environment.[30] Thus, the reason behind this paradox, where common goods are treated as if they were nobody's stuff—*res nullius*—lies in the fact that neither the functional object of the planet is defined nor its holders. Therefore, the legal challenge that humanity faces is to transform these intangible "garbage dumps" with "insufficient delimitation of property rights,"[31] into a new object of law that exists autonomously relative to the legal status of the territories where the great circulation systems operate. Only by defining the good itself and the precise legal effects emerging from the relationships generated around it, to be possible to build a *Collective human action.*[32]

10.2 Functional Aspect Versus Static Territorial Sovereignty of the Earth: Historical Dysfunctionality

The possibility of abandoning an exclusively territorial approach to the planet, by recognizing a complementary approach that deals with the previously ignored *functional aspect*, would allow the application of different legal regimes to the different components of the Earth System, taking as criteria their different essences—*qualitative/functional vs territorial/static*.

The first attempt to legally represent the functional aspect of the planet took place in 1967, when the Maltese Ambassador to the UN, Arvid Pardo, proposed at the UNGA[33] the initial concept of the Common Heritage of Humankind (CHH),[34] which, by 1971, evolved into a proposal for an ocean treaty in which he tried to demonstrate "how the concept of common heritage could be implemented in the marine environment as a whole."[35] At the time, the scientific knowledge needed to measure and define the "marine environment" as a new legal object distinct and autonomous from the territorial place where seawater temporarily circulates, which did not yet exist. The technical and political infeasibility of this project has meant that the CHH has remained confined to the seabed beyond the limits of the exclusive economic zones. In 1988, Malta proposed that the climate be recognized "as part of the Common Heritage of Humankind"[36] reaffirming that the concept, principles, and fundamental objectives of CHH do not align with the leftover territories under state jurisdictions, but rather focus on the functional/qualitative aspect of the planet. The scientific limitations and the fact that climate "is not restricted to the global commons, but also covers areas subject to national jurisdiction,"[37] have led to the issue being circumvented by using different terminology that allows the law to recognize the *problem,* but not the object: "climate change as a Common Concern of Humankind."[38] This *semantic solution* implies that the Stable Climate, as a phenomenon emerging from the functional aspect of the planet, remains in an undefined legal situation, that is, it is still *res nullius*.

The consequences of this fundamental issue were identified as early as 1991 by one of its main mentors, Mostafa Tolba: "*It is very important that the concept of the Common Concern of Humankind be further elaborated to make its content and scope understandable and clear; it is also important to see how this concept can be interpreted in terms of the rights and obligations of States in the process of its* implementation."[39] Even though it remains the legal framework adopted by the Paris Agreement, since its conception it has not been the subject of any further discussion, "to avoid that the Common Concern of Humankind stays as a vague political formula, which could be used to legitimize the lack of concrete actions by simply declaring an environmental concern."[40]

In 1992, the United Nations Framework Convention on Climate Change (UNFCCC),[41] Article 1(3), defined the *Climate System* as "the whole of the atmosphere, hydrosphere, biosphere and lithosphere and their interactions," integrating into a single concept—interactions—the *functional dimension of the planet* and how it is influenced by those tangible components, from which Climate emerges. Paragraph 10.7 defines "reservoir" as "the set of these components of the climate system, where greenhouse gases are stored," uniting functions and components "as a single system at the level of the planet,"[42] constituting a huge step forward in conceptualizing the functional aspect of the planet but without defining precisely an object and consequently without being able to define its legal status.

10.3 The Functional Aspect of the Earth and the Earth System Sciences

In recent decades, Earth System Sciences have deepened their understanding of the functioning of the Earth system as a single whole, which cannot be fully comprehended by analyzing each of its components in isolation. Such a reductionist approach would, for example, miss out on the emergent properties that can only be understood by considering the unity of the Earth System as a single, deeply interconnected system. "Climate, and in particular global average surface temperature, is an emergent property of Earth System dynamics that affects the entire planet."[43] The chemical composition of the atmosphere that was the basis for the emergence of the period of climate stability after the last glaciation, around 11,700 years ago—the Holocene period—was the result of various natural processes, including the profound gas exchanges that life carried out with the atmosphere and the oceans, which gave rise to a highly complex chemical composition of the atmosphere.[44] The recent change in the chemical structure of the atmosphere resulting from human activities, especially since the 1950s, has made a decisive contribution to the increase in global temperature, which in turn has led to a change in the patterns of thermodynamic behavior between the poles and the tropics, "leading to the destabilization of atmospheric circulation patterns and the slowing down of ocean circulation."[45] This "functional aspect" of our planet is now scientifically **definable and measurable** through the so-called *Planetary Boundaries.*[46,47] These limits or boundaries demonstrate a combination of variables, relationships, and parameters that together describe the functioning of the Earth System and the limits to the degradation of these processes.[48] In other words, they allow us to understand the role of the interaction between chemical, biological, and physical processes in maintaining a favorable state of functioning for humanity (i.e., the Holocene), as well as the role of humanity in pushing the functioning of the System out of this stable and desirable state (i.e., the Anthropocene). Within these limits, the system is resilient—that is, it has the capacity to absorb shocks while maintaining its functioning.[49] When these limits are exceeded, the system no longer tends to recover its original "identity," but tends toward a different configuration—the tipping point.[50] In this way, the favorable biogeophysical conditions that correspond to a well-functioning Earth System are defined qualitatively and quantitatively, giving rise to the Safe Operating Space for Humanity.[51] This transforms the analysis of the state of the climate into an indicator of the state of functioning of the entire system, i.e., a *proxy*[52] for the entire Earth System. With the possibility now open by science to translate the state of the functional aspect of the planet into a set of indicators, i.e., translating and defining what its stability or instability is, and describing a qualitative conceptual space that translates the Holocene-like state, it is now possible to conceptualize an autonomous object for the law from which rights and duties can emerge, opening a possibility not considered by Ostrom: "**Is the conventional theory of collective action is the best theory for analysing how to reduce the massive threats of climate change?**"[53]

10.4 Considering the Functional Aspect of the Earth as the Common Heritage of Humankind

"When we refer to the pattern of stable dynamics of the Earth System that corresponds to a Stable Climate, we are not referring to 'matter' or the physical planet, but to the way matter and energy move and circulate around the planet. Matter is always changing through chemical reactions and physical processes—and, in the long term, through biological evolution. But the patterns and rates of these changes, and their interactions to form higher-order structures such as ecosystems, follow well-defined patterns of organization and stability. On a planetary scale, the way matter and energy move around the planet, creating various patterns of atmospheric and oceanic circulation, follows the laws of thermodynamics and result in a Stable Climate. A Stable Global Climate is something that can only be legally classified as an intangible natural good."[54]

Intangible assets are structural pillars in knowledge-based human societies, which include not only human-created assets but also natural phenomena such as geostationary orbit or radio-magnetic frequencies, in the context of Outer Space Law.[55] The pattern of atmospheric and oceanic circulation that emerged in the Holocene and resulted in a Stable Climate is humanity's most precious asset. Taking into account that a Stable Climate is the basis of everything,[56] we consider the possibility of recognizing a well-defined circulation pattern that results in a Stable Climate as a natural intangible legal object on Earth. Given that climate change "*is an obstacle to the harmonious development of the personality and to quality of life,*[57] and that a "subjective right to a stable climate is thus a personality right,"[58] *also recognized by the United Nations Human Rights Council,*[59] *leads us to argue that the combination of these rights to a Stable Climate constitutes a heritage that should belong collectively to all of humanity—A Common Heritage of Humankind.*

These subjective rights of all human beings have a "**heritage meaning**, once everyone is entitled to receive. *(. . .) It's not just a concern or interest of humanity; it's on a higher scale; it's (. . .) an intangible heritage (. . .). It's an idea that also has a legal basis, meaning that it's not something artificial; this means that we can find a political basis, a foundation that allows the climate system to constitute a logically intangible Heritage of Humanity (. . .),*[60]" this can be the most innovative of all new approaches.[61]

If we recognize equal ownership of the functional aspect of the planet, so that among all the holders there can be obligations resulting from the different uses (appropriation of the common good), or rights resulting from the benefits provided in the good that belongs to everyone (provision of the common good), it is possible to envision new economic or governance mechanisms designed to explicitly recognize and reward human activities or natural processes that have positive impacts and contribute to the restoration and maintenance of a well-functioning Earth System with a Stable Climate.

In this sense, taking into account that: (1) the functional aspect of the planet is indivisible and non-appropriable (though depreciable); (2) everyone has the capacity to influence its functioning, either positively or negatively, and it is impossible to exclude anyone from its use, as all depend on it; (3) the existence of a **heritage meaning** that arises from the need to pass on to future generations the biogeophysical conditions essential for maintaining this favorable state; (4) The current legal framework has already recognized the concept of Common Heritage of humankind; (5) the legal regime of Common Heritage of Humankind can be applied not only to defined territorial areas but also to "elements or ***conceptual areas***, on behalf of all, since a commons area is considered to belong to everyone";[62] (6) climate can be considered a *proxy* of the state of functioning of all the Earth System. These considerations allow us to conceptualize a qualitative non-territorial space—the SOSH—as something that should be owned by humanity as a collective good— **A Stable Climate as Common Heritage of Humankind.** In doing so, we materialize Arvid's seminal project, which focuses on the functional/qualitative aspect of the planet. In 2021, the Portuguese Climate Law (Law no. 98/2021, December 31st) enshrined the objective of recognizing **Stable Climate** as a **Common Heritage of Humankind** at the United Nations in Article 15 f. This innovative objective has made Portugal the first country in the world to legally recognize the functional aspect of the planet in an autonomous way from the territory.

10.5 Governing the Functional Aspect of the Earth: The Condominium Model

The project to legally recognize the "functional aspect" of the Earth System as a Common Heritage of Humankind implies a solution that allows this new legal object not to clash with the "static territorial dimension of sovereignty." Instead, it should serve as a complementary instrument that integrates the sovereign interest of each State into the broader framework of the Earth System, whose preservation and stability are common interests shared by all States and all generations. Furthermore, it encourages us to look at obstacles from a new perspective that until now was considered insurmountable:

> The current international gathering of Nation-State cannot afford to recognize the atmosphere and oceans as global commons (using scientific criteria), because to do so would *immediately and irrevocably* destroy land as private property, since land would qualify as a global commons. (. . .) Furthermore, since the entire paradigm of Nation-States is based on the acceptance of bounded excludable territories, it is likely that we cannot construct global commons without *de-constructing* the Nation-State system.[63]

The problem of reconciling apparently opposing interests—*private and common*—is not new in the legal sciences. It has been organized/harmonized through a private law figure that defines the situation in which a materially indivisible asset belongs to several owners, in which each has private or exclusive property rights over certain fractions while simultaneously sharing ownership of the parts/systems of the building that constitute its structure and common functions. This legal figure is known as a **condominium**.

The success of this legal figure is due precisely to the fact that it carries out an operation of legal division, which is not just related to physical spaces. **The great condominium's innovation** was that, in addition to the division of spaces, it provided for the existence of a division based on functional criteria, identifying structures and systems for common use that are essential to the security and full enjoyment of private properties. This double operation results in an overlap between private and common property, **and from what is apparently antagonistic and irreconcilable, a perfectly symbiotic relationship of harmonizing interdependencies emerges.** The analogy with the Earth and the overlap between its functional aspect and the static territorial concept of sovereignty is perfect.

As far as our planet is concerned, it is not only made up of territory and matter, but it also contains a system that can operate in different ways, resulting in stable or unstable circulation patterns, which in turn can result in a stable or unstable climate. This functional aspect can now be defined, measured, and quantified in nine essential processes.[64] Just like in a condominium, the common systems are identified and shared, regardless of whether they pass through common areas or private fractions. Therefore, a governance system with shared contributions is established to ensure the permanent maintenance of these systems on which all private units depend for the enjoyment of their property rights. In a condominium, no fraction owner sees the common management of the systems as a threat to their property. On the contrary, without ensuring these services of common interest, their ownership cannot be fully exercised. Rather than being a threat, it is a precondition for their full enjoyment. On a global scale, through the legal autonomization of the intangible system, a new object of global governance—The Earth System—could be defined, while still maintaining national territorial sovereignty.[65] Based on this legal support, an accounting system of the different positive and negative contributions can be built, and a process of restoration and permanent maintenance of the common system could be initiated through the Earth Condominium.[66]

10.6 Shifting to Stabilized Pathway: The *Flows* Versus *Stock* and Its Cascade Effects

A human action that actively "entails stewardship of the entire Earth System—biosphere, climate, and societies—and could include decarbonization of the global economy, enhancement of biosphere carbon sinks,"[67] implies assigning economic value to ecosystem services and negative emissions. In fact, the goal of carbon neutrality is an attempt to internalize the negative externalities, to mitigate further environmental degradation. However, with the concentration of CO_2 in the atmosphere beyond the safety threshold and close to an irreversible *tipping point*, it is also necessary to internalize the positive externalities into the economy, that is, to pay for the production of negative emissions.

Due to the Paris Agreement decision to maintain the legal framework of recognizing ***climate change*** as a Common Concern of Mankind,[68] the climate system itself is still in an undefined legal situation, and the system and its reservoir (*stock*) remain a *res nullius*. The result of this option is that, presently, providing benefits in this climate system (*stock*) does not generate a right to compensation (internalization of positive externalities), nor does causing damages generate obligations (internalization of negative externalities). Given this context, the axis of action was to focus the strategy on attempting to ***emissions*** (*flows*),[69] which limits climate action to a game of reduction and neutralizing emission *flows*. From a *Stock-based* perspective this option can be negative or, in the best-case scenario, neutral, which does not guarantee the reduction of CO_2 levels in the atmosphere to avoid a tipping point. This approach omitted the need for the "Provision of Public Goods" giving rise to some pernicious cascade effects, as it is impossible to fulfill the second structural condition to avoid the tragedy of commons: congruence between the norms of provision (reducing the excess *stock*) and those of appropriation of the common good (emissions, *flows*).[70]

From a *flow-based* perspective, reducing emissions means generating a benefit for everyone, and therefore, corresponds to a credit. From the point of view of a *stock* perspective, reducing emissions does not automatically halt the accumulation of CO_2 in the atmosphere and oceans, *and is not enough "to steer the Earth System away from a potential threshold and stabilize it in a habitable interglacial-like state*'.[71] To do so, human societies must also address existing carbon *stocks* and explore strategies to incentivize carbon uptake. When the Paris Agreement states that the objective is to "promote financial *flows* consistent with a low greenhouse gas emissions pathway,"[72] it implicitly recognizes a new activity as creating value in the economy, namely the reduction/mitigation/neutralization of CO_2 emissions. However, this option has the pernicious side effect of perpetuating the association between the creation of value in the economy and the existence of new CO_2 emissions, without considering the creation of value per se that the provision of ecosystem services and negative

CO_2 emissions has for society. "But incentives induce individuals, businesses, and nations to keep on emitting greenhouse gases".[73] In essence, this perpetuates an economic dynamic that relies on value creation from carbon *flows*, while the true wealth creation that ecosystem services and the uptake of CO_2 represent for societies, neglects incentives to address the *stock* of CO_2 already accumulated in the atmosphere and oceans. As Ostrom very well explains:

> *Problems of appropriation are centred on the allocation of flow, problems of supply concern stock. Appropriation problems are independent of time; supply problems depend on time. (. . .) Both problems are always present in the management of a common good to a greater or lesser degree, and therefore the solutions to one problem must be congruent with the solutions to the other.*[74]

Without establishing an economic dynamic focused on *stock* reduction through restoring nature and cleaning up the atmosphere and oceans, there will be no change in the dynamics of emissions production (*flows*), and emissions will continue to rise, or, at least, they will not decrease on the scale that is necessary, because it is antithetical to the very dynamics of the economy.

In addition, this problem of addressing *flows* without addressing *stocks* turns many well-intentioned policies into policies with perverse results, in which boomerang effects and paradoxes take over. Without a policy to intervene in the *stock*, all incremental improvements, even the technological solutions to reduce or remove emissions, may even have perverse effects: we already have a solution to remove; therefore, we can continue to emit, and even increase.[75] These facts are already recognized by the *European Scientific Advisory Board on Climate Change, which* states that removing CO_2 from the atmosphere "poses a dilemma for policymakers, as they need to find ways to incentivize sustainable increases in carbon removal, while avoiding the risk of discourage reduction in greenhouse gas emissions."[76] Changing the current undefined legal status of the "garbage dumps of humanity" is a structural condition for a *human-created stabilized Earth pathway.*[77]

10.7 The Rational Choice

Bearing in mind that there is no longer any credible pathway to 1.5 degrees Celsius,[78] and that the current projections based on NDCs,[79] the unmitigated scenario must also be considered[80]—*Climate-system feedback loops* can generate a chain of changes that are self-reinforcing and could eventually lead to major climate warming.[81] This compels financial experts to reassess economic risks: "Our economy depends on nature. So destroying nature means destroying the economy."[82]

This means that even the logic of the "***theory of* collective *inaction***"[83] can be questioned: the private costs of each one can start to be greater than the costs of cooperating. "Macroeconomic damages from climate change are six times larger than previously thought (. . .) with Social Cost of Carbon of $1,056 per ton of carbon dioxide imply that unilateral decarbonization policy is cost-effective for large countries such as the United States."[84] Taking into account this new relationship between private costs and collective costs, a new perspective of common interest and national interest can be built: assuming private costs, even when the benefits are diffuse, can be cost-effective. This also means restoring the common good with a system of internalizing the benefits, corresponding to the interests of all, becoming the only rational choice. Since the system is global and ecosystem restoration is valid in any territory, there is enormous space and flexibility in ecosystem restoration to carry out negative emissions, creating potential ground for cooperation.

Given the global latitude of the problem, it doesn't matter where these negative emissions can be made because the common system is the same. Given the difference in income between countries, the different contributions over time to the problem, the different types and states of conservation of ecosystems, and their different locations, the prospects for cooperation can be enormous.

In this sense, the countries that have historically have been the biggest emitters of CO_2 for environmental, health, economic, and financial reasons have a vested interest in carrying out a global ecosystem restoration plan that produces negative emissions, creating an economy for the provision of the public good. The countries that have contributed the least to these emissions also share the same interest, as a Stable Climate does not depend only on GEE, but "relies on a resilient and well-functioning biosphere." Taking into account that a significant part of the key ecosystems is found in the countries that have contributed the least to these emissions, there could be a realistic mutual interest between both parts—a large-scale nature restoration project associated with payment for ecosystem services can be the rational choice. The recognition of the functional aspect of the Planet—Climate as a Common Heritage of Humankind—should serve as the legal framework to internalize these current "externalities," creating the proper legal framework to enable collective action.

10.8 Conclusion—The Only Realistic Project

The distinction between the functional aspect of the planet and the static territorial concept of sovereignty put forward by the ILC allows us to bring the functional aspect of the Earth to International Law and Institutions, and has created the possibility of overcoming some structural obstacles that have prevented the possibility of concerted action: (1) *the definition of functional/non-territorial global commons; (2) the possibility of recognizing global commons without deconstructing the Nation-State system, but rather integrating it symbiotically with the Common Heritage of Humankind; (3) the application of the conventional doctrine of collective action to climate change through the condominium approach; (4) the internalization of the vital life-supporting factors that currently constitute externalities, enhancing the emergence of a nature-restoration economy capable of generating the negative feedbacks* to create and maintain a "Stabilized Earth"; *(5)* a theory of justice based on the principle that different contributions to the problem over time should give rise to different contributions to restoration efforts, in the interest of all; *(6)* translate to a legal concept like the Common Heritage of Humankind the fact that ecosystem restoration is valid in any territory, and therefore carrying out negative emissions can be made anywhere. This common legal support creates a vast common ground for cooperation, where differences can be harmonized and channeled into collective action. *However,* for this to happen, the most basic condition for any human enterprise must be in place: a proper legal environment.

Notes

1 United Nations Framework Convention on Climate Change, Rio de Janeiro, June 13, 1992.

2 Paris Agreement (2015). COP21, at https://unfccc.int/process-and-meetings/the-paris-agreement/the-paris-agreement.

3 Energy Transitions Commission. Mind the Gap: How Carbon Dioxide Removals Must Complement Deep Decarbonisation to Keep 1.5°C Alive, p.8. 2022. https://www.energy-transitions.org/publications/mind-the-gapcdr/.

4 Steffen W., Rockström J., Richardson K., Schellnhuber H. Trajectories of the Earth System in the Anthropocene, Edited by William C. Clark, Harvard University, Cambridge, MA, 115 (33) 8252–8259, 2018. https://doi.org/10.1073/pnas.1810141115.

5 Idem—"Stability landscape showing the path of the Earth System out of the Holocene and thus out of the glacial-interglacial boundary cycle to its current position in the warmer Anthropocene. Shown here are the two divergent paths of the Earth System in the future (broken arrows). Currently, the Earth System is on the Greenhouse Earth pathway driven by human GHG emissions and biosphere degradation toward a planetary threshold of ~2 °C, beyond which the system follows an essentially irreversible pathway driven by intrinsic biogeophysical feedbacks. The other path leads to the Stabilized Earth, a path of Earth System management guided by man-made feedbacks toward a quasi-stable, man-maintained basin of attraction. 'Stability' (vertical axis) is defined here as the inverse of the system's potential energy. Systems in a highly stable state (deep valley) have low potential energy and it takes considerable energy to move them out of this stable state. Systems in an unstable state (top of a hill) have high potential energy and require only a little additional energy to push them off the hill and down into a valley of lower potential energy."

6 Lenton, T. M. et al., Climate tipping points—Too risky to bet against. Nature 575, 592–595, 2019.

7 Steffen W., Rockström J., Richardson K., Schellnhuber H. Trajectories of the Earth System in the Anthropocene. William C. Clark, Harvard University, Cambridge, MA, 115 (33) 8252–8259. 2018. https://doi.org/10.1073/pnas.1810141115.

8 UNITED NATIONS ENVIRONMENTAL PROGRAMME (UNEP)—Emissions Gap Report 2024. Consulted in: November 25, 2024. Available at: https://www.unep.org/resources/emissions-gap-report-2024.

9 O'Gorman, Maebh, Global Warming: A Tragedy of the Commons Comparative Research in Law & Political Economy. Research Paper No. 32/2010. http://digitalcommons.osgoode.yorku.ca/clpe/99.

10 Ansari S., Wijeb F., Gray B. Constructing a Climate Change Logic: An Institution Perspective on "Tragedy of the Commons" Organization Science, Vol.24, No.4.. 2013.

11 Bruns, B.- Transforming Climate Dilemmas From Tragedy to Cooperation, Commons Research Innovations, Sheila R. Foster, Chrystie F. Swiney, Cambridge University Press. 2021, https://www.cambridge.org/core/books/abs/cambridge-handbook-of-commons-research-innovations/transforming-climate-dilemmas-from-tragedy-to-cooperation/F48F5E13F57DAB8AEF5987E1637CF3DE.

12 Energy & Climate Intelligence Unit, Negative emissions: why, what, how?, 2018.

13 Olson, M. (1965). The Logic of Collective Action: Public Goods and the Theory of Groups. Cambridge, MA: Harvard University Press.

14 Ostrom, E. (2009). A Polycentric Approach for Coping with Climate Change, Policy Research Working Paper 5095, Background Paper to the 2010 World Development

Report. The World Bank, Development Economics, Office of the Senior Vice President, at https://ssrn.com/abstract=1494833.

[14] Ostrom, E. (2009). A Polycentric Approach for Coping with Climate Change, Policy Research Working Paper 5095, Background Paper to the 2010 World Development Report. The World Bank, Development Economics, Office of the Senior Vice President, in https://ssrn.com/abstract=1494833

15 Hardin, G. (1968). The Tragedy Of The Commons, Science, 162(3859), Pp. 1243–1248.

16 Magalhães, P. Common interest, concern or heritage, The commons as a structural support for a Earth system Law, Chapter 14, Earth System Law, Standing on the precipice od the Anthropocene, Edited by Timothy Cadman, Margot Hulbert and Andrea C. Simonelli, Routledge, 2021.

17 MILUN, K. The political uncommons: The cross-cultural Logic of the Global Commons. Routledge, 2018.

18 Hartzog, Paul B. *Global Commons: Is a definition possible? Department of Political Science, University of Utah. 2003. https://www.academia.edu/2757037/Global_Commons_Is_Definition_Possiblesafeguarding.*

19 Idem 17.

20 Rockstrom J., et al. The planetary commons: A new paradigm for safeguarding Earth- regulating systems in the Anthropocene. PNAS, 2024. https://doi.org/10.1073/pnas.2301531121.

21 Idem 19.

22 Bilial,A & Kanzig,D. The MacroEconomic Impact of Climate Change: Global VS. Local Temperature, National Bureau of Economic Research, Cambridge, 2024.

23 Camilleri, Joseph A. & Jim Falk. *The End of Sovereignty? The Politics of a Shrinking and Fragmenting World.* Aldershot: Edward Elgar Publishers, 1992.

24 MILUN, K. The political uncommons: The cross-cultural Logic of the Global Commons. Routledge, 2018.

25 Vogler, J. The Global Commons: A Regime Analysis. Second edition, Wiley, 1995.

26 Magalhães, P., Steffen, W., Aragão, A., Soromenho-Marques, V. From the Cultural Immaterial towards the Natural Intangible, 2013. https://www.researchgate.net/publication/373237266_From_the_Cultural_Immaterial_towards_the_Natural_Intangible?_tp=eyJjb250ZXh0Ijp7ImZpcnN0UGFnZSI6InByb2ZpbGUCDIJwYWdlIjoicHJvZmlsZSJ9fQ.

27 *A/76/10 Report of* The International Law Commission, United Nations, seventy-second session (26 April- 4 June and 5 July- August 6, 2021), p. 29. Available at: https://legal.un.org/ilc/reports/2021/english/a_76_10_advance.pdf. Accessed on May 7, 2024.

28 Kiss, A. La notion de patrimoine commun de l'humanité, Académie de droit international de La Haye, Recueil des cours, 1982, tome 175, pp. 103-256.

29 Idem7.

30 https://www.economist.com/international/2019/07/18/space-law-is-inadequate-for-the-boom-in-human-activity-there

31 Hardin, G. The Tragedy Of The Commons, Science, 162(3859), 1968, Pp. 1243–1248.

32 Ostrom, E., et al. Revisiting the Commons: Local Lessons, Global Challenges. Science 284 (5412). May 1999. 278–282. https://www.researchgate.net/publication/13102089_Revisiting_the_Commons_Local_Lessons_Global_Challenges.

33 Pardo, A. Document A/6695—Address by Arvid Pardo to the 22th Session of the General Assembly of the United Nations. Official Records of the General Assembly. Twenty-Second Session. Agenda Item 92, 1967..

34 Idem 32.

35 Pardo, A. The Common Heritage of Mankind: Selected Papers on Oceans and World order 1967–1974. 1975. Malta University Press. X-XI.

36 A/43/241 United Nations General Assembly, September 12, 1988.https://digitallibrary.un.org/record/46039.

37 Borg, S.. Climate Change as a Common Concern of Humankind, Twenty Years Later . . . From UNGA to UNSC. Towards an Integrated Climate Change and Energy Policy in the European Union. IUCN Academy of Environmental Law, University of Malta, 2007. http://www.iucnael.org.

38 A/43/905 United Nations General Assembly. November 30, 1988. https://digitallibrary.un.org/search?f1=author&as=1&sf=title&so=a&rm=&m1=p&p1=UN.+General+Assembly+%2843rd+sess.+%3A+1988-1989%29.+2nd+Committee&ln=en.

39 M. Tolba. The Implications of the "Common Concern of Mankind Concept in Global Environmental Issues, IIDH Journal, 13, 1991. pp. 237–246.

40 Idem 38.

41 United Nations Framework Convention on Climate Change, Rio de Janeiro, June 13, 1992.

42 Will Steffen, The Planetary Boundaries Framework: Defining a Safe Operating Space for Humanity, SOS treaty—The safe operating space treaty, A new approach to managing the use of the earth system. Cambridge Scholars Publishing, Cambridge, UK, 2016, p. 43.

43 Steffen, W.; Morgan, J. From the Paris Agreement to the Anthropocene and Planetary Boundaries Framework: an interview with Will Steffen. ISSN 1474–7731, Globalizations, 2021. Pp 1–13. Available in: https://doi.org/10.1080/14747731.2021.1940070. Access in March 14th 2024.

44 Lenton, T.. Earth System Science, A Very Short Introduction, Oxford University Press, 2016, p.18..

45 Magalhães, P., Steffen, W. Why we need a critical legal innovation to save our climate, 2021. https://www.commonhomeofhumanity.org/climate..

46 Rockström, J., Steffen, W., Noone, K. et al.. A safe operating space for humanity, 472–475, Nature, 2009., https://doi.org/10.1038/461472a.

47 Steffen W., Richardson K., Rockström J., Cornell S., Fetzer I., Bennett E., Biggs R., Carpenter S., Vries W., De Wit C., Folke C., Gerten D., Heinke J, Mace G., Persson L, Ramanathan V., Reyers B., Sörlin S. Planetary boundaries: Guiding human development on a changing planet, 2015.

48 Lenton, T. M., & van Oijen, M. (2002). Gaia as a complex adaptive system. Philosophical transactions of the Royal Society of London. Series B, Biological sciences, 357(1421), 683–695, in https://doi.org/10.1098/rstb.2001.1014.

49 Folke, C., Carpenter, S., Elmqvist, T., Gunderson, L., Holling, C.s, Walker, B. (2002). Resilience and Sustainable Development: Building Adaptive Capacity in a World of Transformations. Ambio. 31. 437–40. 10.1579/0044-7447–31.5.437. 2002. https://www.researchgate.net/publication/11086054_Resilience_and_Sustainable_Development_Building_Adaptive_Capacity_in_a_World_of_Transformations.

50 Lenton, T. Earth System Science, A Very Short Introduction, Oxford University Press, 2016. p.18.

51 Rockström, J., Steffen, W., Noone, K. et al. A safe operating space for humanity, 472–475, Nature, 2009. https://doi.org/10.1038/461472a.

52 Magalhães,.P (2024) *The Planet´s "Functional Aspect" and the 1992 Rio Convention:Climate as Nobody's Stuff, or as Common Heritage? CEBRI—Revista do Centro Brasileiro de Relações Intenacionais, Ano 3/Nº10/abr-Jun, P.235-256*

53 Ostrom, E. A Polycentric Approach for Coping with Climate Change, Policy Research Working Paper 5095, Background Paper to the 2010 World Development Report. The World Bank, Development Economics, Office of the Senior Vice President, 2009. https://ssrn.com/abstract=1494833

54 Magalhães, P., Steffen, W. Why we need a critical legal innovation to save our climate. Common Home of Humanity, 2021. Available in: https://www.commonhomeofhumanity.org/climate. Access in March 14th 2024.

55 Since the beginning of the 18th century, the Law has recognized the enormous importance of intangible property as a *fact* that deserves legal recognition, to the extent that intellectual property is now the support of the society of knowledge. In Space Law, intangible legal goods have been extended to natural phenomena such as geostationary orbit or radio magnetic frequencies.

56 Rockstrom, J. Let the environment guide our development, TED Talk, 2010. https://www.ted.com/talks/johan_rockstrom_let_the_environment_guide_our_development .

57 Redinha, R. Guimarães, M.R. (2024) The Case for a Stable Climate, The Urgent Need for a Subjective (Personality?) Right in the Portuguese Legal System, Taylor's and Francis, 2024.
58 Idem 56.
59 Human Rights Council to the environment, 2022. https://www.ohchr.org/en/statements-and-speeches/2022/04/right-healthy-environment.
60 Sobrino, J. M. Património é Uma Ideia (. . .) Património é Algo que é Necessário Conservar no Interesse de Todos. Jornal Quercus, 50 (Jan-Feb), 4–5. 2012. http://www.quercus.pt/images/PDF/QA/QA50.pdf
61 Idem 53.
62 Frakes, Jennifer. The Common Heritage of Mankind Principle and the Deep Seabed, Outer Space, and Antarctica: Will Developed and Developing Nations Reach a Compromise? Wisconsin International Law Journal. 21: 409. 2003.
63 Hartzog, P. B. Global Commons: Is Definition Possible? University of Utah, Department of Political Science. Abril de 2003. https://www.academia.edu/2757037/Global_Commons_Is_Definition_Possible.
64 Rockström, J., Steffen, W., Noone, K. et al. "A safe operating space for humanity", 472–475 , Nature, 2009. https://doi.org/10.1038/461472a.
65 Magalhães P. Common Home of Humanity: White paper. https://www.commonhomeofhumanity.org/_files/ugd/deeae3_b06d317d69ad41739f302d8d2bf95236.pdf
66 Magalhães, P. Earth Condominium – From the Climate Change to a New Juridic Conception of the Planet. Edições Almedina. Coimbra. 2007.
67 Idem7.
68 Paris Agreement - preamble. COP21, 2015.https://unfccc.int/process-and-meetings/the-paris-agreement/the-paris-agreement..
69 Paris Agreement. COP21, 2015.https://unfccc.int/process-and-meetings/the-paris-agreement/the-paris-agreement.
70 Ostrom, E., et al. 1999. Revisiting the Commons: Local Lessons, Global Challenges. Science 284 (5412). Maio de 1999. 278–282. https://www.researchgate.net/publication/13102089_Revisiting_the_Commons_Local_Lessons_Global_Challenges.
71 Idem 7.
72 Paris Agreement Article 2. c). COP21, 2015.https://unfccc.int/process-and-meetings/the-paris-agreement/the-paris-agreement.
73 Bruns, B.Transforming Climate Dilemmas From Tragedy to Cooperation, Commons Research Innovations. Sheila R. Foster, and Chrystie F. Swiney. Cambridge University Press. 2021. https://www.cambridge.org/core/books/abs/cambridge-handbook-of-commons-research-innovations/transforming

-climate-dilemmas-from-tragedy-to-cooperation/F48F5E13F57DAB8AEF5987E1637CF3DE

74 Ostrom, Elinor. Governing the Commons, The evolution of Institutions for Collective Action. James E. Alt, Douglass C. North, The Political Economy of Institutions and Decisions, Cambridge University Press. 1990.

75 MDPI - Carbon Capture and Storage: Application in the Oil and Gas Industry, by Sara Yasemi, Yasin Khalili , Ali Sanati, and Mohammadreza Bagheri. Sustainability 2023, 15(19), 14486. https://doi.org/10.3390/su151914486 Published in 4 October 2023.
https://www.mdpi.com/2071-1050/15/19/14486.

76 European Scientific Advisory Board on Climate Change. Scientific advice for the determination of an EU-wide 2040 climate target and a greenhouse gas budget for 2030–2050. 2023. https://climate-advisory-board.europa.eu/reports-and-publications/scientific-advice-for-the-determination-of-an-eu-wide-2040.

77 Idem 7.

78 UNEP – United Nations Environment Programme. Closing the Window - Emission Gap Report, 2022. https://www.unep.org/resources/emissions-gap-report-2022.

79 Idem 2.

80 *Swiss Re Institute The economics of climate change: no action not an option April 2021*

81 Idem7.

82 Elderson, F. The economy and banks need nature to survive. The ECB Blog, 2023. https://www.ecb.europa.eu/press/blog/date/2023/html/ecb.blog230608~5cffb7c349.en.html#:~:text=Degradation%20of%20nature%20can%20impair,system%20are%20to%20this%20degradation.

83 Ostrom, E. (2009). A Polycentric Approach for Coping with Climate Change, Policy Research Working Paper 5095, Background Paper to the 2010 World Development Report. The World Bank, Development Economics, Office of the Senior Vice President, at https://ssrn.com/abstract=1494833.

84 Bilial,A & Kanzig,D. The Macro Economic Impact of Climate Change: Global VS. Local Temperature, National Bureau of Economic Research, Cambridge. 2024.

11

Climate Change and Sustainable Campus Practices in the Context of Science Diplomacy

Zeynep Olgun, Şefika Şule Erçetin, and Nurşah Ataş

11.1 Introduction

Science plays a significant role in fostering international interactions through its methods of producing, organizing, and documenting knowledge within the framework of ethical values. The strengthening relationship between science and diplomacy contributes to addressing global issues and facilitates the implementation of international scientific collaborations (Turekian et al., 2015). In this context, science diplomacy, which has existed in the past but has gained increasing importance in recent times with the establishment of a more systematic and sustainable structure, not only supports international diplomatic relations but also encourages the global dissemination and effective use of scientific research (Copeland, 2016; Campbell, 2012).

The increase in research and development activities on a global scale, the spread of science and technology production centers worldwide, the strengthening of the understanding that global issues can only be solved through joint scientific efforts, and the rise of consortia and international political and diplomatic participation have played a significant role in the development of science diplomacy (Ruffini, 2017). Science diplomacy is a process that activates scientific cooperation mechanisms to establish constructive international collaborations, manage conflicts, and develop solutions to global issues (Turekian, 2018). Science diplomacy plays an important role in addressing global issues of the twenty-first century, such as food security, poverty reduction, pandemics, mass migrations, and climate change (Galluccio, 2021).

Climate is an area that can be evaluated in many parameters, such as temperature, humidity, and precipitation, on both regional and global scales. In other words, climate is the average of meteorological events such as temperature, air pressure, humidity, wind, and precipitation (Kekül, 2023). In addition to experiencing natural changes within its own system, climate is also influenced by human activities directed toward nature, which have been ongoing for centuries (Açıkalın et al., 2024).

Climate change is the effect of differences and unusual events that emerge as a result of observing average climate values over specific periods, and it impacts ecosystems. Factors such as temperature increases, extreme weather events, rising water levels, and the increased acid levels in oceans due to climate change place living ecosystems at risk (Eren, 2021). Addressing global issues such as environmental problems and climate change, or mitigating their effects, requires the effective conduct of international cooperation and negotiations (Gluckman et al., 2017). In this context, the need for science diplomacy practices is increasingly growing to address the climate crisis at the international level and develop solutions.

The concept of sustainability, aimed at reducing the impacts of climate change and environmental effects, plays an important role in the design of university campuses worldwide. In this context, sustainable campuses are defined as campuses that demonstrate sensitivity to the ecological environment and aim to minimize their negative effects on the environment (Saygın & Ulusoy, 2011). Issues such as energy efficiency, water and waste management, green buildings and infrastructure, sustainable transportation, food, education, and research, which are addressed internationally to reduce the impacts of climate change, constitute the key elements of sustainable campuses (Sonetti et al., 2016).

The aim of the study is to evaluate climate change and sustainable campus practices within the framework of science diplomacy. To achieve this, the concepts of science diplomacy, climate change, and sustainable campuses are first explained, followed by a discussion of climate change and sustainable campus practices within the context of science diplomacy.

11.2 The Concept and Dimensions of Science Diplomacy

Today, with the increasing need for more scientific data in areas such as climate change, pandemics, migration, security issues, and warfare, policymakers have begun to strengthen international interactions and develop strategies and policies (Ruffini, 2017). At this point, the evolving relationship between science and diplomacy has contributed to the resolution of international issues and supported the implementation of scientific collaborations (Turekian et al., 2015).

Science diplomacy has been a challenging concept to define by politicians, diplomats, and scientists (Flink & Rüffin, 2019). However, when examining the literature, science diplomacy is expressed as the use of scientific collaborations to address global issues, create constructive international partnerships, and manage conflicts (Kıran & Açıkalın, 2021). In addition to promoting scientific cooperation, science diplomacy is also seen as a tool for countries to achieve their national interests and goals on the international stage (Ruffini, 2017). In this context, science diplomacy brings together scientists, diplomats, and decision-makers based on the principles of knowledge sharing and collaboration (Turekian, 2018).

The concept of science diplomacy has evolved over time with different dimensions, tools, and practices. Especially in recent years, the increase in international collaborations in the fields of science and technology, the sharing of scientific innovations, and the dissemination of inventions have contributed to the development of science diplomacy (Kulalı Martin, 2021). With the advancement of digital technologies, scientific work conducted by universities, non-governmental organizations, international companies, and science and technology centers has added new dimensions to science diplomacy (Fedoroff, 2009).

In 2009, the Royal Society in the United Kingdom, in collaboration with AAAS, organized a meeting on "New Frontiers in Science Diplomacy," which was attended by government ministers, scientists, diplomats, policymakers, business leaders, and journalists. During this meeting, science diplomacy was classified into three main categories: science in diplomacy, diplomacy for science, and science for diplomacy. This classification has been used in both academic and theoretical studies worldwide, as well as in practical applications in the field of international relations (The Royal Society, 2010).

Science in Diplomacy refers to the production of scientific knowledge and its effective use in international negotiations and decision-making processes. Just as important as the production of scientific data is the ability of diplomats to understand and utilize this information (Milkoreit, 2015). The collaboration between organizations leading international scientific efforts and foreign ministries significantly enhances countries' capacities to achieve their foreign policy goals (The Royal Society, 2010). There are many international institutions and organizations that contribute to the development of the concept of science in diplomacy. For example, the Intergovernmental Science-Policy Platform on Biodiversity and Ecosystem Services (IPBES), established by ninety-four governments in Panama City in 2012, works to strengthen the science-policy relationship in the areas of biodiversity conservation, sustainability, and ecosystem studies (Kulalı Martin, 2021). Another example of science for diplomacy is the Antalya Diplomacy Forum, which has been held annually in Türkiye since 2021. This forum facilitates high-level meetings between professionals such as heads of state, political leaders, diplomats, academics, experts, and businesspeople, enabling the exchange of ideas on diplomacy, policy, regional and global issues, and addressing international challenges (Şehitoğlu & Eren, 2022).

Diplomacy for Science refers to the use of diplomatic means to enhance international scientific collaborations. Today, many countries enter into bilateral agreements and hold summits at the governmental level to facilitate scientific research and secure joint funding (Dehgan & Colglazier, 2012; The Royal Society, 2010). For example, government officials, diplomats, and scientists from various countries participate in global efforts to address issues such as climate change, food and energy security, and pandemics, creating international networks and working toward common goals (Fedoroff, 2009). One of the best examples of diplomacy for science is the Intergovernmental Panel on Climate Change (IPCC). Established in 1988 by the World Meteorological Organization (WMO) and the United Nations Environment Programme (UNEP), the IPCC was created to provide scientific assessments on the current state of climate change and its potential environmental and socioeconomic consequences. The IPCC conducts periodic evaluations of the latest research in scientific, technical, and socioeconomic fields. As a result, scientists from around the world volunteer to contribute to these efforts (Gluckman et al., 2017; Yıldırım, 2019).

Science for Diplomacy involves the establishment of scientific collaborations between countries to improve relations, using scientific activities and platforms to form partnerships. When traditional diplomatic channels fall short, fostering scientific collaborations allows for the restoration of relationships between states (Kıran & Açıkalın, 2021; The Royal Society, 2010). The efforts of scientists aimed at reducing political tensions between countries, the use of credible and effective scientific data in negotiation and mediation processes, and platforms that emphasize the universality of science and common cultural interests are examples of the concept of science for diplomacy (Caymaz & Özsoy, 2022; Flink & Rüffin, 2019).

11.3 Science Diplomacy and Climate Change

11.3.1 Climate Change in Historical Process

Climate change has emerged as a result of profound changes in human lifestyles and global production and manufacturing management. As such, climate change has become a central issue in the decision-making mechanisms of states from economic, social, and political perspectives. Climate change has not only affected environmental issues but also directly impacted the development strategies of countries, as well as their policies on security, energy, health, and agriculture (Kekül, 2023; Tuğaç, 2022).

Throughout the 4.5 billion years of Earth's history, changes in the climate system have been observed many times (Şen, 2022), and these changes have been interpreted in different ways over various historical periods. For instance, in the 1200s, a mild yet variable climate type prevailed. Natural disasters such as storms, floods, heavy rainfall, and droughts occurred. Between the 1500s and 1800s, the "Little Ice Age" took place, a period characterized by an increase in harsh, cold, and rainy days. Due to abnormal climatic conditions, many people and creatures, particularly in Europe and other regions, perished from freezing (Öztürk, 2002). In the nineteenth century, scientists focused on the Ice Age and argued that an increase in temperature could prevent the threat of glaciers and that the production of CO_2 and the consumption of fossil fuels would be beneficial for humanity (Kekül, 2023).

With the impact of the Industrial Revolution, there has been an increase in temperature values in both the Northern and Southern Hemispheres. After industrialization, global issues such as sudden temperature changes, forest fires, rising sea levels, glacial melting, and drought have emerged (Von-Storch & Stehr, 2006). Climate change was addressed at the United Nations (UN) Environment Conference and the Stockholm Declaration in 1972, emphasizing that global warming is harmful to the planet and needs to be fought (Poorhashemi, 2024). Furthermore, according to the World Meteorological Organization and the UN Office for Disaster Risk Reduction (UNDRR) report, half of the disasters that occurred between 1970 and 2019 were caused by weather, climate, and water-related events. These disasters resulted in the loss of 2 million lives and an economic loss of 3.64 trillion dollars (Akıllı, 2023).

Scientific data have shown that climate change is becoming an increasing threat to living ecosystems and that, besides natural factors, human activities are accelerating changes in the climate system. Research indicates that global temperatures have risen by 0.8 degrees Celsius over the past century, and if serious measures are not taken by countries (as long as CO_2 levels continue to rise), global temperatures could increase by an average of 1.5–4.5 degrees Celsius by 2050 (Demirbaş & Aydın, 2020). At this point, scientists emphasize that global warming must be limited to 1.5 degrees Celsius; otherwise, there will be an increase in extreme weather events, a decrease in water resources, a decline in agricultural production, and severe issues such as migration waves (Ministry of Environment, Urbanization, and Climate Change, 2024).

The issue of climate change is not merely an environmental problem but also demonstrates the existence of global social injustice. For example, developing countries such as Nepal, Afghanistan, Somalia, and Kenya, despite having the least impact on climate change, are among the nations experiencing its effects most severely (Metin, 2023). The fact that climate change affects all living ecosystems on a global scale and leads to social injustices among countries has made collaborative efforts between supranational international organizations and governments a necessity (Levy & Patz, 2015). This situation has paved the way for the organization of a series of international treaties, protocols, agreements, and panels with the participation of states worldwide.

11.4 Science Diplomacy and Climate Change Agreements

In 1985, the Vienna Convention and the Montreal Protocol went down in history as the first protocol to state, based on scientific findings, that human-induced factors causing the depletion of the ozone layer needed to be restricted. This protocol, applied to the ozone layer laid, the foundation of the climate change regime (Goyal et al., 2019). In 1988, at the Intergovernmental Panel on Climate Change (IPCC), scientific research conducted by scientists on the subject was presented to policymakers (Hulme & Mahony, 2010). On the other hand, at the 1992 Rio Conference on Environment and Development, the United Nations Framework Convention on Climate Change (UNFCCC) was signed by 158 countries, discussing measures to be taken to prevent human-induced causes in order to reduce the danger of greenhouse gases to the climate system (Kekül, 2023). In 1997, the Kyoto Protocol, the first implementation agreement of the UNFCCC, imposed obligations on states regarding the reduction of greenhouse gas emissions. However, the countries that committed to implementing these obligations did not assume any legal responsibility. The reason for this was interpreted as the inadequacy of the protocol's scope and the inconsistency of the targets set (Rosen, 2014).

In the 2009 Copenhagen Conference, five key points for international cooperation on climate change were addressed: providing detailed and systematic scientific data on the climate crisis, monitoring international policies and strategies, coordinating implementations, mobilizing technology, and efficiently managing the economy, and implementing a climate strategy trade plan among regions and countries (Busby, 2010). When comparing the Copenhagen Accord with the Kyoto Protocol, the literature reveals both positive and negative assessments. A negative evaluation is that the Copenhagen Accord fell behind the Kyoto Protocol in terms of binding provisions, whereas on the positive side, the United States and China reached an agreement on restricting gas emissions, commitments were made to provide financial and technological support to developing countries, and the impact of deforestation on climate. Under the 2015 Paris Climate Agreement, accepted in 2015, all countries worldwide pledged to reduce greenhouse gas emissions post-2020. This agreement established a framework program addressing national contributions, financing, technology development and transfer, capacity building, negotiation, and communication in the fight against climate change (Ministry of Foreign Affairs, 2023; Kekül, 2023). This program includes many targets such as renewable energy, transitioning to zero carbon emissions, and stabilizing greenhouse gas concentrations. These targets are believed to be beneficial for both the economy and a livable world (Erickson & Brase, 2019).

To mitigate the adverse effects of climate change, it has been crucial for countries such as the United States, the European Union, and China, as well as supranational civil society organizations, to address the issue in all its dimensions, formulate policies and strategies, and develop multilateral partnerships, agreements, forums, and projects. These initiatives have also influenced the efforts of other nations (Kekül, 2023). In this context, climate change, which negatively affects habitats on a global scale, has become one of the most critical issues of the twenty-first century and has been discussed across interdisciplinary fields such as international relations, political science, economics, health, and education.

The problems identified within the framework of climate change must be resolved through international cooperation based on scientifically supported evidence (Ruffini, 2018). In this regard, science diplomacy plays a crucial role in the fight against climate change by integrating scientific data into policy-making processes and developing sustainable international cooperation networks.

A global issue such as environmental problems and climate change can be evaluated within the three main classifications of science diplomacy. The discussion of scientific knowledge in international negotiations through the Intergovernmental Panel on Climate Change reflects the approach of science in diplomacy. The development of international networks through climate change agreements at the state level serves the concept of diplomacy for science, while promoting international scientific collaborations to raise global climate awareness falls within the scope of science for diplomacy. In this respect, climate change emerges as a topic that can be examined in all aspects of science diplomacy (Gluckman et al., 2017; Kökyay et al., 2023)

11.5 Sustainable Development and Sustainable Campuses

In 1972, with the United Nations (UN) organized Human and Environment Conference (Stockholm Declaration), climate change and environmental issues were addressed at a global level, and the concept of sustainability emerged. Sustainability focuses on creating long-term positive impacts of human activities on the environment, economy, and society. In this context, the UN's 1987 publication of the Our Common Future (Brundtland) Report combined the concept of sustainability with development policies and introduced the concept of sustainable development, aimed at meeting the needs of the present without compromising the ability of future generations to meet their own needs (Handl, 2012; Wright, 2002).

Conceptually, sustainable development has attracted the attention of policymakers, non-governmental organizations, and academics, becoming a key priority within countries' policy and strategy areas such as economy, health, security, and education (Yılmaz Fındık & Erçetin, 2023). In this regard, the 2015 UN Sustainable Development Goals include the reduction of poverty and hunger, ensuring healthy and quality lives, quality education, gender equality, clean water and sanitation, healthy lives on land and in water, affordable and clean energy, decent work and economic growth, industry, innovation and infrastructure, reduction of inequalities, sustainable cities and communities, responsible consumption and production, and climate action (United Nations Turkey, 2024).

Higher education institutions play a significant role in achieving the Sustainable Development Goals due to their population size and the activities they carry out (Pardal et al., 2020). In line with the Sustainable Development

Goals, higher education institutions are working in various areas to increase environmental awareness and ensure the fair use of resources. One such initiative is sustainable campuses. Sustainable campuses are educational institutions that adopt environmentally friendly practices, promote the efficient use of resources, and aim to raise social awareness. These campuses aim to reduce negative environmental impacts by implementing sustainable practices (Sugiarto et al., 2022). Sustainable campus practices can be categorized as follows:

- Energy and climate change,
- Water, food, and waste management,
- Physical infrastructure and green buildings,
- Transportation,
- Education and research (Kayapınar Kaya et al., 2019).

Sustainable campuses developed within the framework of the Sustainable Development Goals play a critical role in ensuring energy efficiency and supporting environmental sustainability. In this context, reducing the use of fossil fuels, utilizing renewable energy sources, and maintaining infrastructure efficiency are some of the practices implemented in sustainable campuses (Chowdhury et al., 2018; Faghihi et al., 2015). Sustainable campuses help minimize the carbon footprint by reducing the use of natural resources and play an important role in combating climate change (Pandey et al., 2015; Zhu et al., 2018).

Another area of focus in sustainable campuses is reducing water consumption and conserving water resources. In this regard, sustainable campuses with ecological-oriented policies implement measures such as water-saving devices, rainwater harvesting systems, and the recycling of wastewater. These practices can help reduce water consumption on campuses (Peng et al., 2018). Another significant practice in sustainable campuses is promoting the consumption of local and organic food, using sustainable agricultural products in campus cafeterias, recycling food waste through composting, and conducting awareness campaigns to prevent food waste (Barlett, 2011). Waste management initiatives can positively influence the community's attitudes toward waste management and waste reduction awareness. In this context, sustainable campuses that successfully implement waste management see an increase in recycling rates. Additionally, waste reduction is achieved through waste management strategies implemented via composting center programs (Smyth et al., 2010). In collaboration with non-governmental organizations, municipalities, and public institutions, sustainable campuses use recyclable materials and organize educational programs on recycling to engage in community service activities (Sugiarto et al., 2022).

One of the other important practices of sustainable campuses is physical planning and ecological priorities. In this context, sustainable campuses implement green building designs that increase energy and water efficiency, are constructed with environmentally friendly materials, and do not harm human health. The infrastructure systems used in sustainable campuses are also designed to be environmentally conscious and aim for cost savings (Petratos & Damaskou, 2015).

In sustainable campuses, practices such as greening transportation areas, promoting the use of electric vehicles and public transportation, and implementing bicycle-sharing systems aim to reduce the carbon footprint caused by transportation (Atherton & Giurco, 2011). In this context, designing environmentally friendly transportation vehicles, planning efficient transportation management systems, and collaborating with industrial organizations to develop supportive infrastructure and facilities are among the practices of sustainable campuses. Moreover, the efficient transportation systems developed by sustainable campuses can serve as examples for the general transportation system of the city in which they are located (Bond & Steiner, 2006).

Sustainable campuses promote sustainability not only through physical infrastructure practices but also through educational and research activities (Wimala et al., 2016). Environmentally friendly seminars, conferences, and research projects are conducted for students and staff. Additionally, to contribute to the development of environmentally conscious generations, sustainable environmental education is included in university curricula (Hays & Reinders, 2020). These efforts help increase environmental awareness in society and contribute to the adoption of sustainable lifestyles.

11.6 Sustainable Campus Practices in Science Diplomacy

Sustainable campuses are increasing the leadership role of higher education institutions in providing solutions to global environmental issues. Higher education institutions actively engaged in sustainability practices can inspire policymakers, public institutions, and non-governmental organizations, and have a significant influence on both national and international public opinion (Sugiarto et al., 2022).

In this context, there is a direct relationship between the sustainable campus practices of higher education institutions and science diplomacy. Sustainable campus initiatives carried out within the scope of science diplomacy gain greater visibility on an international level and facilitate the development of stronger global collaborations. At the same time, addressing environmental issues and spreading scientific knowledge within the framework of sustainable campus practices contributes to science diplomacy. Sustainable campuses raise awareness by providing educational and capacity-building opportunities for students from foreign countries and help bring together the knowledge and experiences of different nations.

11.7 Conclusion

Science diplomacy is an essential tool for countries to achieve their national interests and goals on the international stage and to promote scientific collaboration in addressing global issues. In the twenty-first century, global challenges such as increased migration, pandemics, natural disasters, environmental issues, and climate change have played a significant role in the effectiveness of science diplomacy. For example, addressing issues related to climate change, such as greenhouse gas emissions, temperature rise, and rising sea levels, with scientific data and information on both national and international platforms, has facilitated the creation of the Paris Climate Change Agreement and contributed to governments' formulation of climate policies. Moreover, the transfer of green energy and environmentally friendly technologies from developed countries to developing nations, as well as increasing these countries' capacity to fight climate change, is made possible through science diplomacy. On the other hand, science diplomacy practices can bring together scientists from different countries to develop a shared scientific understanding and language on climate change. This helps reduce differences in views between states and facilitates the identification of common goals. For instance, the scientific data presented by the Intergovernmental Panel on Climate Change (IPCC), which is widely supported by scientists around the world, has laid the foundation for global climate agreements. In summary, science diplomacy plays a critical role in creating more effective, inclusive, and sustainable solutions by combining scientific knowledge and international cooperation in the fight against climate change.

Sustainability is a concept that encompasses science, society, and the economy, with the aim of having a positive long-term impact on all living beings. Issues related to climate change and the environment, which are part of the United Nations (UN) sustainable development goals, have been addressed with both economic and social dimensions. In this context, combating climate change globally, protecting biodiversity, and ensuring social justice, political stability, and economic development are seen as essential imperatives. Climate change is an issue that needs to be addressed not only through emission reductions but also through technological and sustainable nature-based global solutions.

Sustainable campuses are considered an application area in the fight against climate change. Within the framework of the UN Sustainable Development Goals, innovative projects are being developed in sustainable campuses on issues such as energy, water, food and waste management, and climate change. Through scientific diplomacy, these projects are being disseminated internationally and presented as applicable models in foreign countries. Additionally, within the scope of sustainable campus practices, scientific research is being conducted on topics such as climate change, energy efficiency, and environmental management, providing policymakers with scientific data. These efforts can be influential in shaping the foreign policy goals of states within the framework of scientific diplomacy. Scientific diplomacy plays a crucial role in expanding the innovative technologies developed in sustainable campuses to other countries through bilateral and multilateral agreements between states. On the other hand, within the scope of student and faculty mobility studies, awareness about sustainable campus practices is being promoted.

Reference

Açıkalın, Ş. N., Sarı, E., & Erçetin, Ş. Ş. (2024). Role of education in awareness on climate change. *Current Perspectives in Social Sciences*, *28*(1), 56–63. https://doi.org/10.53487/atasobed.1454546

Akıllı, E. (2023, December). COP28 İklim Zirvesi ve Türkiye'nin Küresel İklim Değişikliği ile Mücadelesi. *SETA Yayınları.* https://www.setav.org/assets/uploads/2023/12/P379.pdf

Atherton, A., & Giurco, D. (2011). Campus sustainability: Climate change, transport and paper reduction. *International Journal of Sustainability in Higher Education, 12*(3), 269–279. https://doi.org/10.1108/14676371111148054

Barlett, Peggy F. (2011). Campus sustainable food projects: Critique and engagement. *American Anthropologist, 113*(1), 101–115. https://doi.org/10.1111/j.1548-1433.2010.01309.x. (2011).

Bond, A., & Steiner, R. (2006). Sustainable campus transportation through transit partnership and transportation demand management: A case study from the university of florida. *Berkeley Planning Journal, 19(*1), 126–142. https://doi.org/10.5070/BP319111492.

Busby, Joshua W. (2010). *After Copenhagen: Climate governance and the road ahead.* Council on Foreign Relations. http://www.jstor.org/stable/resrep00274

Campbell, C. (2012). A consortium model for science engagement: Lessons from the U.S.-DPRK experience. *Science & Diplomacy, 1*(2), 1–9.

Caymaz, E., & Özsoy, B. (2022). Türkiye'nin Kutup Bilim Diplomasisi. *İletişim ve Diplomasi 7*, 5–24. https://doi.org/10.54722/iletisimvediplomasi.1098563

Chowdhury, N., Hossain, C. A., Longo, M., & Yaïci, W. (2018). Optimization of solar energy system for the electric vehicle at university campus in Dhaka, Bangladesh. *Energies, 11*(9), 2433. https://doi.org/10.3390/en11092433.

Copeland, D. (2016). Science diplomacy. In C. M. Constantinou, P. Kerr, & P. Sharp (Eds.), *The SAGE handbook of diplomacy.* SAGE Publications.

Çevre, Şehircilik ve İklim Bakanlığı. (2024, December). İklim Değişikliği. https://iklim.gov.tr/sss/iklim-degisikligi

Dehgan, A., & Colglazier E. W. (2012) Development science and science diplomacy. *Science & Diplomacy, 1*(4), 1–6.

Demirbaş, M., & Rozelin, A. (2020). 21. Yüzyılın En Büyük Tehdidi: Küresel İklim Değişikliği. *Ecological Life Sciences, 15*(4), 163–79.

Eren, O. E. (2021). *The role of global climate change negotiations in the people's republic of China - United States competition* (Master's Thesis, Yalova University).

Erickson, L. E., & Brase, G. (2019). Paris Agreement on climate change. In *Reducing greenhouse gas emissions and improving air quality.* CRC Press.

Faghihi, V., Hessami, A. R., & Ford, D. N. (2015). Sustainable campus ïmprovement program design using energy efficiency and conservation. *Journal of Cleaner Production, 107*(16), 400–409. https://doi.org/10.1016/j.jclepro.2014.12.040

Fedoroff, N. V. (2009). Science diplomacy in the 21st century. *Cell Magazine, 136*(1), 9–11. https://doi.org/10.1016/j.cell.2008.12.030

Flink, T., & Rüffin, N. (2019). The current state of the art of science diplomacy. In D. Simon, S. Kuhlmann, J. Stamm, & W. Canzler (Eds.), *Handbook on science and public policy*. Edward Elgar. https://doi.org/10.4337/9781784715946

Galluccio, M. (2021). *Science and diplomacy Negotiating essential alliances.* Springer.

Gluckman, P. D., Turekian, V. C., Grimes, R. W., & Kishi, T. (2017). Science diplomacy: A pragmatic perspective from the inside. *Science & Diplomacy*, *6*(4), 1–13.

Goyal, R., England, M. H., Gupta, A. S., & Jucker, M. (2019) Reduction in surface climate change achieved by the 1987 montreal protocol. *Environmental Research Letters, 14*(12), 1–10. https://doi.org/10.1088/1748-9326/ab4874

Handl, G. (2012) Declaration of the United Nations Conference on the Human Environment (Stockholm Declaration), 1972 and the Rio Declaration on Environment and Development, 1992. *United Nations Audiovisual Library of International Law, 11*(6), 1–11.

Hays, J., & Reinders, H. (2020). Sustainable learning and education: A curriculum for the future. *International Review of Education*, *66*, 29–52. https://doi.org/10.1007/s11159-020-09820-7

Hulme, M., & Mahony, M. (2010). Climate change: What do we know about the IPCC? *Progress in Physical Geography: Earth and Environment*, *34*(5), 705–718. https://doi.org/10.1177/030913331037371

Kayapınar Kaya, S., Dal, M., & Aşkın, A. (2019). Türkiye'deki Devlet ve Vakıf Üniversite Kampüslerinin Sürdürülebilir-Ekolojik Parametreleri Açısından Karşılaştırılması. *Balıkesir Üniversitesi Fen Bilimleri Enstitüsü Dergisi, 21*(1), 106–125. https://doi.org/10.25092/baunfbed.532420

Kekül, M. B. (2023). *Evaluation of EU-Turkey relations in the context of climate diplomacy*. (Master's thesis, İstanbul Bahçeşehir University).

Kıran, Y. S., & Açıkalın, Ş. N. (2021). New tools of soft power: Turkey's education and science diplomacy. *Hacettepe Üniversitesi Eğitim Fakültesi Dergisi*, *36*(4), 977–985. https://doi.org/10.16986/HUJE.2021072717

Kökyay, F., Uysal, A. C., & Kökyay, N. E. (2023). Bilim Diplomasisi Bağlamında Ortak Araştırma İstasyonlarının Türkiye'nin Antarktika Vizyonuna Katkısı. *Nişantaşı Üniversitesi Sosyal Bilimler Dergisi, 11*(1), 232–250. https://doi.org/10.52122/nisantasisbd.1259713

Kulalı Martin, Y. (2021). İskandinavya'nın Bilim Diplomasisindeki Rolü: Nobel Ödülleri Örneği. *Galatasaray Üniversitesi İletişim Dergisi*, *6*, 57–74. https://doi.org/10.16878/gsuilet.867918

Levy, B. S., & Patz, J. A. (2015). climate change, human rights, and social justice. *Annals of Global Health*, *81*(3), 310–322. https://doi.org/10.1016/j.aogh.2015.08.008

Metin, S. (2023). Afrika'da İklim Değişikliğinin Etkileri: İklim Adaleti ve Göç. *Akdeniz Havzası ve Afrika Medeniyetleri Dergisi*, *5*(2), 99–118. https://doi.org/10.54132/akaf.1402787

Milkoreit, M. (2015). Science and climate change diplomacy: Cognitive limits and the need to reinvent science communication. In Lloyd S. Davis and Robert G. Patman (Eds.), *Science diplomacy new day or false dawn?* World Scientific. https://doi.org/10.1142/9789814440073_0006

Öztürk, K. (2002). Küresel İklim Değişikliği ve Türkiye'ye Olası Etkileri. *Gazi Eğitim Fakültesi Dergisi*, *22*(1), 47–65.

Pandey, A., Shiva Shankar, Y., Saxena, A., and Khandelwal, R. (2015). Harnessing solar energy for sustainable energy management ïn the campus: Case study of JUET, Guna. *History*, *41*(190), 156–162.

Pardal, A., Romeira, T., & Durão, A. (2020). Eco green campus: Challenges and opportunities. The study case of polytechnic institute of Beja. Proc. *E3S Web of Confenrance*, *171*, 1–8. https://www.e3s-conferences.org/articles/e3sconf/pdf/2020/31/e3sconf_eepgtech2019_01010.pdf

Peng, S., Cui, H., & Ji, M. (2018). Sustainable rainwater utilization and water circulation model for green campus design at Tianjin University. *Journal of Sustainable Water in the Built Environment*, *4*(1). https://doi.org/10.1061/JSWBAY.000084.

Petratos, P., & Damaskou, E. (2015). Management strategies for sustainability education, planning, design, energy conservation ïn california higher education. *International Journal of Sustainability in Higher Education*, *16*(4), 576–603. https://doi.org/10.1108/IJSHE-03-2014-0038

Poorhashemi, A. (2024). International environmental law. In Sué González Hauck, Raffaela Kunz, and M. Milas (Eds.), *Public international law: A multi-perspective approach*. Routledge. https://doi.org/10.4324/9781003451327–19

Republic of Turkey Ministry of Foreign Affairs. (2023). *Paris Anlaşması [Paris Agreement]*. June 16, 2023. https://www.mfa.gov.tr/paris-anlasmasi.tr.mfa.

Rosen, A. M. (2014). The wrong solution at the right time: The failure of the kyoto protocol on climate change. *Politics & Policy*, *43*(1), 30–58. https://doi.org/10.1111/polp.12105

Ruffini, P. B. (2017). *Science and diplomacy a new dimension of ïnternational relations.* Springer.

Ruffini, P. B. (2018). The Intergovernmental Panel on Climate Change and the science-diplomacy nexus. *Global Policy*, *9*(3), 73–77. https://doi.org/10.1111/1758–5899.12588

Saygın, N., & Ulusoy, P. (2011). Sürdürülebilir Kampüs Tasarımı için Yağmursuyu Yönetimi ve Yeşil Altyapı Teknikleri. *Politeknik Dergisi*, *14*(3), 223–231.

Smyth, D. P., Fredeen, A. L., & Booth, A. L. (2010). Reducing solid waste ïn higher education: The first step towards 'greening'a university campus. *Resources, Conservation and Recycling*, *54*(11), 1007–1016. https://doi.org/10.1016/j.resconrec.2010.02.008

Sonetti, G., Patrizia Lombardi, L., & Chelleri, L. (2016). True green and sustainable university campuses? Toward a clusters approach. *Sustainability*, *8*(1), 83. https://doi.org/10.3390/su8010083

Sugiarto, A., Lee, C.-W., & Huruta, A. D. (2022). A systematic review of the sustainable campus concept. *Behavioral Sciences*, *12*(5), 130. https://doi.org/10.3390/bs12050130

Şehitoğlu, R., & Eren, E. (2022). Konferans Diplomasisi Perspektifinde Antalya Diplomasi Forumu'nun Uluslararası İlişkilerdeki Rolü. *İletişim ve Diplomasi*, *7*, 185–204. https://doi.org/10.54722/iletisimvediplomasi.1111491

Şen, Z. (2022). İklim Değişikliği ve Türkiye. *Çevre Şehir ve İklim Dergisi*, *1*(1), 1–19.

The Royal Society. (2010). *New frontiers in science diplomacy. Navigating the changing balance of power*. https://royalsociety.org/-/media/policy/publications/2010/4294969468.pdf

Tuğaç, Ç. (2022). İklim Değişikliği Krizi ve Şehirler. *Çevre Şehir ve İklim Dergisi*, *1*(1), 38–60.

Turekian, V. (2018). The evolution of science diplomacy. *Global Policy*, *9*,(3), 5–7. https://doi.org/10.1111/1758-5899.12622

Turekian, V. C., Macindoe, S., Copeland, D., Davis, L. S., Patman, R. G., & Pozza, M. (2015). The emergence of science diplomacy. In Lloyd S. Davis & Robert G. Patman (Eds.), *Science Diplomacy New Day Or False Dawn?* World Scientific. https://doi.org/10.1142/9789814440073_0001

United Nations Turkey. (2024). *Türkiye Sürdürülebilir Kalkınma Amaçları çalışmalarımız*. https://turkiye .un .org /tr /sdgs.

Von Storch, H., & Stehr, N. (2006). Anthropogenic climate change: A reason for concern since the 18th century and earlier. *Geografiska Annaler: Series A, Physical Geography*, *88*, 107–113. https://doi.org/10.1111/j.0435-3676.2006.00288.x

Wimala, M., Akmalah, E., Irawati, I., & Sururi, M. R. (2016). Overcoming the obstacles to green campus implementation in indonesia. *International Journal of Civil and Environmental Engineering*, *10*(10), 1360–1365.

Wright, T. S. (2002). Definitions and frameworks for environmental sustainability in higher education. *Higher Education Policy*, *15*(2), 105–120. https://doi.org/10.1016/S0952-8733(02)00002-8

Yıldırım, S. (2019). *Applicability of science diplomacy and commitments of Paris Agreement in realist theory: Turkey case*. Master's Thesis, İstanbul University.

Yılmaz Fındık, L., & Erçetin, Ş. Ş. (2023). How do universities in türkiye integrate sustainable development goals into their strategies? *Sustainability*, *15*(24), 1–16. https://doi.org/10.3390/su152416799

Zhu, Y., Wang, F., & Yan, J. (2018). The potential of distributed energy resources in building sustainable campus: The case of Sichuan University. *Energy Procedia*, *145*, 582–585. https://doi.org/10.1016/j.egypro.2018.04.085

12

Disinformation and Post-Truth

A Barrier for Climate Diplomacy

Majda Tafra-Vlahović *President of HINA Croatian News Agency, CROATIA*

Climate change is one of the most pressing global challenges, requiring coordinated international action. However, the rise of disinformation and post-truth politics presents a significant obstacle to effective climate diplomacy. Why is that? And how is it that the deliberate spread of false or misleading information undermines global efforts to address the climate crisis? These are the questions this chapter will try to deal with, the aim is to open ways for new questions.

Climate-related disinformation manifests in multiple ways. It is often through deliberate strategies designed to mislead the public, manipulate policy debates, and erode confidence in the scientific consensus on climate change. All of which are extremely dangerous.

One of the most notable examples dates back to the 1980s and 1990s, when ExxonMobil and other fossil fuel corporations funded research and public campaigns aimed at downplaying climate risks. Internal documents reveal that ExxonMobil's scientists privately acknowledged the reality of climate change as early as the 1970s. A 2023 study published in *Science* analyzed these documents, demonstrating that between 1977 and 2003, ExxonMobil's researchers made highly accurate predictions about global warming trends (Supran et al., 2023). Despite this, the company engaged in public campaigns that sowed doubt about climate science. Why? They wanted to protect their corporate interests. Further investigative reports by *Inside Climate News* uncovered how ExxonMobil conducted extensive climate research but later spearheaded efforts to deny the scientific consensus that its own scientists had confirmed. This deliberate strategy not only misled the public but also contributed to decades of policy inaction.

Climate-related disinformation can be categorized into misinformation (inadvertent spreading of incorrect information) and disinformation (deliberate efforts to deceive for political or financial gain). The distinction is crucial, as both forms influence public perception and policymaking.

Fossil fuel companies have historically been key players in disinformation efforts. For example, the American Petroleum Institute (API) funded campaigns framing climate change as an unproven theory rather than an established fact. This strategy mirrors tactics used by the tobacco industry, which for decades funded research questioning the link between smoking and cancer to delay regulatory action (Oreskes & Conway, 2010). Similar methods—emphasizing uncertainty, funding contrarian voices, and promoting misleading narratives—have been deployed to stall climate policy.

One of the most insidious forms of disinformation is climate delayism, which does not deny climate change altogether but rather promotes narratives advocating for postponing action. This tactic is often employed by fossil fuel companies and their allies to protect economic interests. A striking example is BP's introduction of the concept of an individual carbon footprint in 2004, a marketing effort that shifted climate responsibility from corporations to individuals. By focusing public attention on personal consumption rather than systemic emissions from major polluters, this narrative effectively delayed meaningful regulatory action.

Recent research suggests that climate delayism is now the dominant form of climate misinformation. A report by the Center for Countering Digital Hate (CCDH) analyzed over 12,000 YouTube videos from 2018 to 2023, revealing that by 2023, 70 percent of climate denial content had transitioned to promoting delay narratives (Allgaier, 2019; Bounegru et al., 2020; Torrico, 2023). These narratives often emphasize the alleged economic downsides of transitioning to renewable energy, presenting non-transformative solutions while overstating job losses and energy costs. This shift away from outright denial to strategic delayism represents an evolving challenge in climate communication.

The rapid spread of climate disinformation has been exacerbated by social media algorithms, which prioritize sensational or emotionally charged content. Research indicates that false or misleading climate claims are significantly more likely to be reshared than accurate scientific information (Graham et al., 2015).

During the Covid-19 pandemic, climate disinformation became intertwined with broader cultural and political narratives, such as anti-vaccine sentiment and critiques of "wokeness." This blending of misinformation not only polarizes climate discourse but also creates ideological roadblocks to climate action. Astroturfing—the practice of establishing seemingly grassroots organizations that oppose climate policies—further compounds the issue by giving the illusion of widespread public skepticism.

O'Connor & Weatherall (2019) analyze how misinformation proliferates through social networks, emphasizing that false beliefs persist not due to ignorance alone but because of social trust and credibility mechanisms. Echo chambers reinforce climate skepticism, making individuals more susceptible to disinformation that aligns with their pre-existing beliefs.

The impact of climate disinformation is profound, undermining international climate agreements, delaying policy action, and fueling ideological polarization.

The Intergovernmental Panel on Climate Change (IPCC) has consistently been affirming that climate change is real, human-induced, and an urgent global threat. However, coordinated disinformation campaigns have sown doubt about these findings, reducing public urgency to act. Farrell (2016) highlights how corporate funding has significantly contributed to ideological divides, weakening international agreements like the Paris Accord by fostering public skepticism and eroding political will for cooperative solutions.

Disinformation also leads to policy paralysis. Fossil fuel interests played a role in influencing the US withdrawal from the Paris Agreement in 2017 by spreading misleading claims about job losses and economic downturns. Despite evidence that transitioning to a green economy could create millions of jobs, these narratives provided political cover for leaders seeking to dilute climate commitments.

Addressing climate disinformation requires a multi-pronged approach. Strengthening media literacy, promoting fact-checking initiatives, and enforcing platform accountability for spreading false narratives are critical steps. Policymakers must also implement transparency measures to track corporate influence on climate policy. As climate misinformation tactics continue to evolve, recognizing and countering these narratives is essential for fostering informed public discourse and advancing effective climate diplomacy. The challenge is not just to combat denial but to dismantle the strategic delays that hinder necessary action in the fight against climate change.

Efforts to counteract disinformation have focused on multiple strategies: Fact-Checking Initiatives—Organizations like Climate Feedback and Media Matters work to debunk false climate claims, providing accurate information to the public; Media Literacy Campaigns—Enhancing public understanding of how to critically evaluate sources helps counter the spread of misinformation; Regulatory Measures—Governments and international organizations are introducing legislation mandating transparency in climate-related political advertising and corporate communications; Corporate Accountability—Public and shareholder pressure is forcing some fossil fuel companies to alter their narratives, though critics argue that much of this amounts to *greenwashing;* Disinformation has fueled political polarization, complicating efforts to build consensus on climate action. In many countries, climate change has become a divisive issue, with individuals aligning their beliefs along partisan lines.

For instance, in the United States, conservative media outlets had frequently promoted skepticism about climate science, contributing to stark divisions in public opinion. Research suggests that while most Democrats have supported aggressive climate policies, a significant proportion of Republicans have remained skeptical, impeding bipartisan efforts to implement effective legislation (McCright & Dunlap, 2011; Guber et al., 2020). This political divide has hindered climate-related bills, not due to scientific evidence but because of ideological resistance.

Disinformation has also been used as a tool in resource conflicts, particularly concerning transboundary water disputes. In regions such as the Middle East and North Africa, where climate change is intensifying water scarcity, false narratives about water usage have heightened diplomatic tensions. Disinformation that distorts facts about water resource management can create distrust, complicating cooperative efforts to address climate-related challenges.

Social media platforms have amplified climate disinformation through engagement-driven algorithms that prioritize sensational content. Although companies like Meta (Facebook), Twitter, and YouTube have implemented fact-checking initiatives, critics argue these measures lack consistency and transparency. Some key challenges include: Inconsistent policies—While Pinterest has robust definitions of climate disinformation, YouTube and Twitter lack comprehensive guidelines; Limited transparency—Platforms often do not disclose the results of their enforcement actions, making it difficult to assess effectiveness; Resource constraints—Many tech companies underinvest in fact-checking and content moderation, leaving disinformation unchecked.

The fight against climate disinformation is a battle for the future of our planet. The spread of false narratives, driven by corporate interests and political agendas, has delayed critical climate action, undermined scientific consensus, and fostered polarization. Tackling this issue requires a comprehensive approach, including strengthening media literacy, enforcing accountability, and promoting collaboration between governments, tech companies, and civil society. As disinformation tactics continue to evolve, vigilance is essential to ensure that the integrity of climate science and diplomacy is upheld. The stakes could not be higher—confronting this challenge is imperative for securing a sustainable future for generations to come.

References

Allgaier, J. (2019). Science and cnvironmental communication on YouTube: Strategically distorted communications in online videos on climate change and climate engineering. *Frontiers, 4*. https://doi.org/10.3389/fcomm.2019.00036

Bounegru, L., Pryck, K. D., Venturini, T., & Mauri, M. (2020). We only have 12 years: YouTube and the IPCC report on global warming of 1.5°C. *First Monday*, 25(2), 1.

Farrell, J. (2016). Corporate funding and ideological polarization about climate change. *Proceedings of the National Academy of Sciences, 113*(1), 92–97.

Graham, M. W., Avery, E. J., & Park, S. (2015). The role of social media in climate change issue attention and opinion formation. *Journal of Public Affairs, 15*(2), 145–156.

Guber, D., Bohr, J., & Dunlap, R. (2020). 'Time to wake up': Climate change advocacy in a polarized Congress, 1996–2015. *Environmental Politics, 29*(6), 1094–1120. https: // doi.org/10.1080/09644016.2020.1786333.

McCright, A., & Dunlap, R. (2011). The politicization of climate change and polarization in the American public's views of global warming, 2001–2010. *Sociological Quarterly, 52*(2), 155–194.

O'Connor, C., & Weatherall, J. O. (2019). *The misinformation age: How false beliefs spread*. Yale University Press.

Oreskes, N., & Conway, E. M. (2010). *Merchants of doubt: How a handful of scientists obscured the truth on issues from tobacco smoke to global warming*. Bloomsbury Press

Supran, G., Oreskes, N., & Rahmstorf, S. (2023). Assessing ExxonMobil's global warming projections. *Science, 379*(6628), eabk0063. https:10.1126/science.abk0063

Torrico, D. V. (2023). Analysis of climate denialism on YouTube: Refuting instead of debating. *Catalan Journal of Communication & Cultural Studies, 15*(2), 217–235..

13

Political Progress on International and National Climate Diplomacy

Taedong Lee[1], Dian Wahyu Utami

13.1 Introduction

Climate change represents a global crisis that is impacting many sectors beyond national borders. The concept of "climate change" is integrated into international agreements, bureaucratic frameworks, and educational materials (Woodward, 2019) due to the exponential increase of the effects felt by many sectors (Huq et al., 2007; Ibrahim & Johansson, 2022; Leal Filho, 2022). The term has gained prominence alongside the rising greenhouse gas emissions and the associated issues, particularly during the past 400 years (Woodward, 2019). This situation leads to an increase in extreme weather events, such as droughts, storms, and flooding, which jeopardize food security (Singh & Singh, 2019), contribute to rampant biodiversity loss (UNDP, 2024), and exacerbate global health issues (Agache et al., 2021; St. Louis & Hess, 2008). The impacts of these phenomena will continue to worsen without the implementation of effective strategies (UNDP, 2024).

Following the effects of climate change that have become prevalent, the issue of climate change is no longer just an environmental issue that can be addressed by some people or groups only. Climate change has evolved into a complex, multidimensional challenge that touches every aspect of human life, wherein the acts of one nation may result in consequences experienced by others (SITASI). This issue is intensified by the disproportionate allocation of responsibilities and resources among states, coupled with the social, economic, political, and geographical conditions of each nation (Samson et al., 2011). Therefore, the climate change issue has become a significant global political concern that requires comprehensive handling, involving many sectors, people, groups, and nations globally. The transboundary impacts of climate change necessitate international collaboration, as no single nation can successfully handle the issue independently.

The necessity for comprehensive handling and international collaboration has led to major political progress in climate diplomacy. While climate diplomacy creates the framework for collaboration and accountability at the international level, the political progress builds the foundation for countries to actively engage in global efforts. The political progress aims to tackle the collective challenges posed by climate change by forcing countries and international organizations to work together. As climate change escalates, political progress in climate diplomacy becomes increasingly crucial, providing the basis for collaborative actions that combine national interests with the global requirement to address climate change. This progress has been characterized by significant agreements, such as the Kyoto Protocol and the Paris Agreement, which have been able to inspire countries around the world to participate in climate change discussions (Dimitrov, 2019). This cooperative approach is crucial for conserving ecosystems, promoting fair adaptation, and ensuring a sustainable future for future generations.

This book chapter reviews the rapidly growing literature on political progress in climate diplomacy, aiming to identify what factors contribute to successful outcomes and barriers in that matter. The political progress both at the international and national levels will be explored to give a more comprehensive context.Moreover, by presenting practical insight, this chapter will elaborate more on how governance and leadership reforms might transcend challenges and promote significant advancement in global climate diplomacy. Previous research on political progress in climate diplomacy has focused largely on policy frameworks and international agreements, such as the Paris Agreement and Kyoto Protocol, which outline ambitious climate goals. However, much of this literature has overlooked the nuanced factors that drive successful political progress at both international and national levels. Key studies often emphasize high-level policy targets without thoroughly analyzing the underlying mechanisms—such as governance structures, leadership commitment, and public engagement—that enable or hinder these goals. Additionally, many researchers have concentrated on isolated barriers like economic dependencies or political polarization, but have not sufficiently explored how these obstacles can be addressed through integrated governance and leadership reforms. This narrow focus on either policy output or individual barriers has led to an incomplete understanding of political progress in climate diplomacy, lacking practical guidance on how diverse governance models and reform strategies can support effective climate action. By addressing these gaps, this research aims to provide a more holistic analysis of the factors contributing to successful political progress while also offering solutions to overcome persistent barriers. This approach is essential for developing adaptable, context-sensitive strategies that can be applied across different governance levels to achieve meaningful climate outcomes.

This chapter begins with the understanding of what political progress means and how political progress can succeed in negating climate diplomacy (section 13.2). However, some of the key challenges are described in section 13.3 . Thus, it leads to the assessment of the role of climate governance and leadership in tackling the barrier (section 13.4). Moreover, in section 13.5, the case study on how climate diplomacy is shaping at a country level will be explored. Finally, some conclusions based on patterns identified in this chapter will be shown (section 13.6).

13.2 Political Progress in Climate Diplomacy: Literature Review

Political progress in climate diplomacy refers to the advancements made within political systems, institutions, and policies that effectively address climate challenges at both national and international levels. It encompasses the capacity of governments and leaders to enact meaningful policies, respond to scientific findings, and engage the public in achieving climate objectives. Unlike short-term political gains, political progress in climate diplomacy is a sustained effort to adapt governance structures, strengthen leadership, and mobilize collective action toward ambitious climate goals.

13.2.1 Polycentric Governance

Ostrom (2010) introduces polycentric governance, which refers to a framework in which several governing entities function at multiple levels with overlapping jurisdictions in tackling climate change. In the discussion of climate change, polycentric governance is important since it promotes multiple actors in the decision-making process on climate change-related policies (Jordan et al., 2018).

Stripple & Bulkeley (2013), in their work "Governing the Climate," highlight the importance of polycentric governance in addressing climate change, characterized by multiple, overlapping governance structures functioning at local, national, and international levels. Polycentric systems promote innovation and responsiveness through collaboration across various governance scales, enhancing the effectiveness of climate initiatives in addressing mitigation and adaptation efforts (Ostrom, 2010). Hulme (2009) in "Why We Disagree on Climate Change" serves as a foundation for understanding how countries have argued, presented, and addressed climate change issues in markedly different ways. Hulme contends that climate change transcends mere scientific or environmental concerns, emerging as a cultural, political, and ethical phenomenon that provokes diverse interpretations rooted in values, beliefs, and worldviews (Hulme, 2010). Consequently, each nation adopts a distinct method of addressing climate change. This is further backed by nations' geopolitical and geographical positions, which significantly influence the impact of climate change they experience. This decentralized approach facilitates flexible and adaptive climate policies while engaging a broader spectrum of stakeholders, including governments, businesses, and civil society actors. Therefore, climate politics should be understood as a dynamic interaction between domestic and international levels (Tosun & Peters, 2021). This proved that the political progress in climate diplomacy issues is not a one-way process; instead, it is multiple ways and needs multiple perspectives and approaches to national and international political progress.

13.2.2 Leadership

Leadership in addressing climate change remains a crucial area of study since political leaders and actors continue to shape policy and action at both the national and international levels (Gaora et al., 2023). At the national level, leadership in climate diplomacy often manifests through policy innovation and the advocacy of climate initiatives. Leaders in certain nations at any level can set priorities, mobilize resources, and inspire collaborative actions to address the climate issue within the countries and beyond. For example, the change of president in the United States has had a major impact on climate policy, for example, when the Biden administration came to power in 2021 and officially rejoined the Paris Agreement after four years of US disengagement and denial on climate issues (Hyeonjung, 2021). This, of course, changes all US policies, both domestic and foreign. Moreover, the United Kingdom is also showing leadership action by being the first country to pass a legally enforceable Climate Change Act (2008), pledging to an 80 percent reduction in emissions by 2050 (Climate Change Act 2008, Vezirgiannidou, 2015). This illustrates how national leadership can establish standards for other nations. Another instance is the situation in the developing nation of Indonesia, under President Jokowi's second term from 2019 to 2024, during which Indonesia's climate diplomacy aims to reconcile national interests with global collaboration, leading to extensive multilateral cooperation to speed up the energy transition and mitigate carbon emissions (Gaora et al., 2023).

Leadership in climate diplomacy is not only limited to the role of a figure or political leader in a country. But further than that, institutional leadership is very important in creating frameworks and mechanisms that can serve as a global basis (Dimitrov et al., 2019; Purvis & Stevenson, 2010). Institutions such as the UNFCCC create legally binding commitments, establishing a global agreement (i.e., limiting global warming to 1.5 degrees Celsius–2 degrees Celsius), and promote accountability through systems for data sharing and transparency (Kinley et al., 2021). In addition, the actions taken by Pacific island nations have also shown substantial leadership in climate diplomacy (Morgan et al., 2024). Pacific islands face a huge impact from climate change even though they contribute minimally to global emissions; therefore, they are active in advocating for climate change issues on the global stage by forming the Alliance of Small Island States (AOSIS), despite their limited economic and political power (Morgan et al., 2024; Nunn, 2009; Taylor & Kumar, 2016). Another example of how leadership institutions in climate diplomacy look like is the one in the European Union (EU), which aspires to global leadership by influencing international environmental policy addressing climate change. Through several policies, including the European Green Deal, the EU succeeded in integrating economic and social policies to incorporate climate goals into larger policy areas, "organizational façades" (Çelik, 2022). This strategy enables the EU to advocate for climate initiatives, promote collaboration, and build a climate financial mechanism for developing countries (ibid). Thus, leadership in climate diplomacy can be undertaken by anyone and is important for reconciling national interests with global priorities, promoting new policies, and encouraging collaborative efforts to address climate change.

13.2.3 Public Policy

Public policy refers to decisions made by the government in the form of regulations, laws, actions, and priorities to address all the issues arising within the country. According to the classic political scientist, David Easton (1953), public policy is an output of the continuous political system (Easton 1953;Williams & Easton, 1954). Thomas R Dye (1984) defined the public policy as "anything a government choose to do or not to do." Therefore, it can be interpreted that public policy is a policy taken by the government for the public interest at all levels of government, whether local, regional, national, or international. In the issue of climate diplomacy, this public policy plays an important role in realizing global or international commitments into a practical and actionable framework in each country. Effective public policies synchronize international targets, such as those specified in the Paris Agreement, with national priorities, allowing nations to tackle the complex difficulties presented by climate change. Policies such as renewable energy incentives, carbon pricing, and climate adaptation plans exemplify how public policy fosters collaboration among governments, the private sector, and civil society. However, public policy often also hinders climate diplomacy efforts when there is no integration across all levels of governance, especially in countries that adopt a decentralized system. (Urwin & Jordan, 2008). Consequently, integration necessitates the incorporation of climate policy, including the alignment of international agreements with local frameworks. Furthermore, increasing public awareness and encouraging support for climate efforts are vital for guaranteeing policy effectiveness (Stoutenborough & Vedlitz, 2014). By promoting coherence among the various levels of government and facilitating public engagement, public policy in climate diplomacy serves as an effective instrument for attaining sustainable and equitable climate solutions.

13.3 Factor Contributing to Successful Political Progress

Political progress in climate diplomacy is determined by a multiplicity of linked factors that work across multiple levels of governance and involve varied actors. Achieving progress in climate diplomacy means that the actions can successfully address the challenges posed by climate change, including the transboundary effects, socioeconomic impacts, and climate commitments. Moreover, the connection between leadership, governance structures, and how they translate into concrete actions is key to the success of climate diplomacy. Therefore, understanding the factors that contribute to political progress across international, national, and non-state levels is crucial for formulating policies to tackle climate change holistically and equitably.

13.3.1 International Level

The international political discourse about climate change emerged a long time ago. Climate change became a focal point in political discourse in 1988 during the Toronto Conference on the Changing Atmosphere (Tosun & Peters, 2021). Subsequently, following the United Nations Framework Convention on Climate Change (UNFCCC) in 1992, countries globally have committed to responding to the effects of climate change (United Nations, 1992), countries globally have been committed to responding to the effects of climate change (UNFCCC, 1992). Furthermore, the parties that ratified the UNFCCC have convened annually since 1995 to address diverse climate change issues and assessments required at the Conference of the Parties (COP) at various venues (Tosun & Peters, 2021). Numerous agreements and strategies to address climate change have been implemented to date. One of the important milestones in the international agreement on climate change was the publication of the 1997 Kyoto Protocols, which aimed to reduce greenhouse gas (GHG) emissions from industrialized nations by 5.2 percent relative to 1,990 levels by the year 2012 (Harris & Lee, 2017).

Furthermore, following the COP21 in 2015, the Paris Agreement has emerged. The 2015 Paris Agreement establishes a framework for global initiatives to mitigate and adapt to climate change by reducing greenhouse gas emissions and holding world average temperature rising below two degrees Celsius (United Nations Framework Convention on Climate Change (UNFCCC), 2015). In accordance with the objectives of the Paris Agreement, nationally determined contributions (NDCs) serve as the foundation for each nation's commitment to mitigating greenhouse gas (GHG) emissions (Carlson et al., 2021; Peterson et al., 2023). Dimitrov (2016) highlights that the Paris Agreement represents a major political success in international climate negotiations, largely because of its inclusive and flexible structure. The agreement managed to bring together countries with divergent interests by allowing for both top-down and bottom-up approaches (ibid). Through the NDCs scheme, the Paris Agreement facilitates national political progress and international collaboration. This feature indicates that progress in climate diplomacy can be attained through systems that provide flexibility while upholding a robust international oversight framework.

In addition to increasing countries' awareness to participate in climate change commitments, the UNFCCC framework has also succeeded in mobilizing increased finance and technology transfer to support emission reduction and move toward a green transition, especially in developing countries. Climate finance mechanisms such as the Green bonds market (Banga, 2019), Green Climate Fund (Bowman & Minas, 2019), and non-market financing mechanisms (Jakob et al., 2015) proven to be able to mobilize billions of dollars to help developing countries with climate adaptation and mitigation. In addition, one of the essential aspects of international climate negotiations is technology transfer. For instance, through Foreign Direct Investment (FDI) and collaboration with international institutions and R&D, China and India have succeeded in integrating global knowledge and domestic capabilities for development in the solar photovoltaic (PV) industry (Fu & Zhang, 2011). Another example is the technology transfer on carbon capture and storage (CCS) technology, which has been developed in developed countries such as Norway and the United States and has been transferred to other countries such as China (H. Liu & Liang, 2011). Thus, the bilateral and multilateral partnerships have also enhanced international climate diplomacy.

At the international level, successful political progress relies heavily on collective action with robust multilateral frameworks, cooperation mechanisms, and equitable agreements. The role of international institutions such as UNFCCC and agreements such as the Kyoto Protocol and Paris Agreement demonstrates how the collective commitment can drive political progress. By sharing the burden and fostering a common understanding of the effects of climate change, which result in variations in socioeconomic conditions across countries, international actions ensure all countries can participate in climate solutions. These agreements have also included instruments like nationally determined contributions (NDCs), carbon markets, climate funding, and technology transfer, which enable countries to adjust their obligations while supporting collective progress (Ryu & Lee, 2024). These achievements underscore the potential of international collaboration, driven by competent leadership and common goals, to yield significant progress in climate diplomacy.

13.3.2 National Level

National-level political progress plays an essential role in climate diplomacy as it allows countries to address climate change through context-specific policies that correspond with their own political, social, and economic conditions. While international frameworks and negotiations set broad targets for climate action, the actual implementation and innovation of climate policies often occur at the national level (Compston, 2010). In realizing national political progress, participation from various parties is needed, and the policy-making process also depends heavily on how national and international actors, institutions, and networks influence it (Tosun & Peters, 2021). This is in line with the polycentric governance model, as discussed before, which states that climate change diplomacy can be achieved through a decentralized approach that involves multiple levels of government and non-state actors (Ostrom, 2010).

In response to increasing global pressure and international agreements on climate change, countries are increasingly framing their climate policies within national borders. Research conducted by Dubash et al. (2013) indicated that between 2007 and 2012, there was an 11 percent rise in the number of nations implementing climate strategies. This indicates that countries formerly reliant on international agreements (Giddens, 2008) have begun implementing internal measures. Furthermore, the data indicates a rise in climate strategies within developing nations (Dubash et al., 2013). This development indicates that governments acknowledge the need to synchronize global climate targets with national priorities. Moreover, strong political leaders play an essential role in driving national climate policies. For instance, the Korean Green New Deal Policy initiative aimed at achieving carbon neutrality and transitioning to a sustainable, low-carbon economy. This policy proposes three main areas: green infrastructure transitions, low-carbon and decentralized energy supply, and supporting innovation in the green industry (Lee & Woo, 2020). Similarly, Germany's Energiewende (energy transition) policy aims to reduce greenhouse gas emissions and transition to renewable energy. This initiative introduces key mechanisms such as the feed-in tariff system, promoting decentralized energy production and incentivizing technological innovation, making renewables increasingly competitive with conventional energy sources (Rechsteiner, 2021).

National political progress is frequently marked by the formulation of novel policies that incorporate climate variables across several sectors. The Bangladesh Cyclone Preparedness Program is an illustration of how the nation confronts particular climate-induced hazards (Islam, 2014). Rather than concentrating on the design of adaptation and mitigation strategies, Bangladesh's climate diplomacy predominantly emphasizes obtaining financial assistance for adaptation and enhancing resilience to climate impacts, given the significant disruptions caused by climate-induced disasters in the country. Similarly, Tuvalu, as a small island developing state (SIDS), faces existential threats from rising sea levels and extreme weather events. Therefore, instead of relying on international migration as a primary solution, the government of Tuvalu's climate diplomacy focuses on advocating for international funding to support local adaptation projects such as enhanced rainwater harvesting systems and coastal protection measures(McCubbin et al., 2015).

A key aspect of the effectiveness of domestic policy, particularly with climate change, is the public's awareness and engagement. Public knowledge and understanding of climate risk and environmental concerns are essential for effective policy formulation and execution. Public engagement will facilitate the implementation of the strategy on preparedness for climate-induced disasters, such as floods, droughts, and heatwaves (Khatibi et al., 2021). In climate policy, particularly in adaptation efforts like constructing new infrastructure and developing more environmentally sustainable housing, community engagement can be crucial. An instance in Australia shows that participatory approaches employed by three local governments can result in more informed and sustainable adaptation programs (Serrao-Neumann et al., 2015). Thus, public engagement promotes the incorporation of diverse viewpoints, ensuring that policies are both feasible and inclusive.

13.3.3 The Role of Non-State Actor

Based on the polycentric governance model by Ostrom (2010), non-state actors, such as businesses, cities, and Non-Governmental Organizations (NGOs), have an important role in driving climate action (Amanatidis & Urbina Trevino, 2018; Ostrom, 2010). Non-state actors are important in filling the governance gaps and often set ambitious targets beyond national governments (Jordan et al., 2018). As Tänzler and Carius (2013) highlighted that climate diplomacy must go beyond state actors and include civil society, which can drive accountability and ambition in climate policy. Moreover, non-state actors, particularly NGOs, often work beyond national borders to advocate for stronger climate commitments. Therefore, it is important to include the role of non-state actors in discussing political progress on climate diplomacy.

13.3.3.1 Non-Governmental Organizations (NGOs)

NGOs have long been at the forefront of advocacy and awareness-raising in climate diplomacy. Following the research on non-state actors in the global climate regime by Raustiala and Bridgeman (2007), NGOs in this context are characterized as structured non-state entities that aim to instigate change, specifically regarding the nature, form, or extent of international, national, and local responses to climate change. The role of NGOs in climate diplomacy manifests in many forms, including gathering public opinions, providing assistance, participating in climate negotiations, and mobilizing public support. Most NGOs that work on climate diplomacy, such as Greenpeace, WWF, and Climate Action Network, have "consultative status" with the United Nations that allows them formal access to UN documents, negotiations, and deliberations related to climate issues (Raustiala & Bridgeman, 2007). Moreover, NGOs often act as "diplomats" who can perform many functions like a country in climate-related policy negotiations (Betsill & Corell, 2007). Several examples show the NGOs' involvement in climate diplomacy. Climate Action Network, for instance, has pushed for more ambitious emission reduction targets and climate justice issues at international climate negotiations (Jordan et al., 2018). Bulkeley and Schroeder (2012) also provide examples of how NGOs in Los Angeles play a role with local government partners in the pursuit of climate governance. Mayaputri et al., (2024) provide an example of how NGOs help to raise awareness and socialize climate justice in Pari Island, Indonesia. Furthermore, the Youth Climate Movement (YCM) NGOs play a role in increasing the awareness of young people in the world to participate in climate campaigns and oversee the implementation of climate norms under the UNFCCC and Paris Agreement (Govindaraj & S, 2023). These examples show that NGOs have a very important role in climate diplomacy at both national and international levels. Thus, the NGOs and community movements are crucial in enhancing public participation, particularly through community initiatives and grassroots movements, fostering a bottom-up momentum for more robust climate pledges (Feola & Nunes, 2014; Smith et al., 2014).

13.3.3.2 Companies and Private Sectors

The private sector is increasingly recognized as a vital player in climate diplomacy (Krabbe et al., 2015; Wang & Sueyoshi, 2018). Although governments are pivotal in establishing energy policies and regulations, companies possess the capacity to effect substantial change through their business practices and investments (Lee & Morgan, 2024). Numerous corporations illustrate how enterprises navigate the energy transition through the utilization of renewable energy, the implementation of sustainable business practices, and advocacy for legislative reforms. Hsu et al. (2015) argue that companies, especially those in the energy and technology sectors, have the potential to drive innovation and contribute to the reduction of greenhouse gas emissions. Many corporations have begun setting their own emissions targets and investing in sustainable technologies, contributing to the overall political progress in climate diplomacy (Wang & Sueyoshi, 2018). Multinational corporations such as Hyundai, for instance, have made significant developments in electric vehicles (EVs) and fuel cell electric vehicles (FCEVs) that play a pivotal role in the Korean government's roadmap for the hydrogen economy (E. Lee & Mah, 2021). Hyundai has also actively participated in Korea's Supply Chain Environmental Management (SCEM) program by providing training, workshops, and technical support to suppliers, enabling them to implement pollution reduction and energy-efficient practices (K. Lee & Cheong, 2011)

In summary, while public institutions achieve political progress in climate diplomacy through policies and regulations, non-state actors advance in climate diplomacy by leveraging advocacy, innovation, and collaboration. Their ability to raise awareness, mobilize public support, and advocate for important issues to be discussed in both national and international forums is one of the keys to their success. Additionally, non-state actors in the form of companies and corporations have sufficient capital to mobilize resources through investment and support climate finance. Their adaptability, autonomy, and ability to consolidate various parties make non-state actors essential in tackling climate issues and enhancing state-led initiatives.

13.4 Barriers to Political Progress and Solutions

Despite all the achievements in climate diplomacy, numerous barriers and challenges hinder political progress both at international and national levels. These challenges generally originate from political, economic, and institutional dynamics that often create ineffective climate policies. Therefore, it is necessary to improve the obstacles that arise through innovative governance approaches, strong leadership, and collaborative efforts.

13.4.1 Barriers to Political Progress

13.4.1.1 Complexity of Multilateral Negotiations

Despite international agreement in climate diplomacy receiving huge participation and nations' engagement and optimism, however, international agreements such as the Paris Agreement have received several criticisms, especially regarding their effectiveness in mitigating climate change. A primary criticism is the non-binding character of the nationally determined contributions (NDCs), which lack enforceable measures, resulting in inconsistent implementation among nations and frequently causing countries to establish aspirational targets without a feasible strategy to achieve them (Mehrotra & Benjamin, 2022; Pauw et al., 2018; Stankovic et al., 2023). Moreover, 195 countries voluntarily submit their compliance targets, but the detailing method, transparency, and review mechanisms are still ineffective (Raiser et al., 2020).

Moreover, while international political progress on climate diplomacy provides a framework for collective global action, another issue that frequently obstructs collaborative global efforts in climate diplomacy is the reassertion of national interests. Kythreotis (2012) contends that national territoriality, characterized by governments prioritizing sovereignty and economic development, often diminishes the efficacy of international agreements. National interests substantially impede collective global efforts in climate diplomacy, chiefly because of the disproportionate allocation of costs and benefits related to climate change. This imbalance compels nations to prioritize their perceived interests over cooperative initiatives, culminating in ineffectual international accords such as the Kyoto Protocol (Cole, 2008). While some countries often engage in international climate diplomacy due to international pressure, vulnerable states such as Small Island Developing States (SIDS) deeply rely on collective action for their existential threats from climate change (Schroeder & Nath, 2021). Consequently, the diversity of priorities and national interests in involvement in establishing climate target ambitions frequently lacks robust national leadership to synchronize domestic policy with international commitments. International frameworks just establish a foundation for collective action; nonetheless, their efficacy is contingent upon the commitments and actions of individual nations.

13.4.1.2 Conflicting National Priorities

Several examples of climate diplomacy show the importance of national progress in both adaptation and mitigation policies in realizing climate diplomacy. However, critics also argue that these national-level climate policies are often difficult to compare between countries and lead to uneven progress (Tosun & Peters, 2021). A further critique of national climate diplomacy is the inconsistency of regulations and practices within the nation itself. For example, Brazil has developed a strategy in climate diplomacy, with the main objectives being forest protection and emission reduction targets. However, political shifts and administrative changes have caused pro-climate policies to be discontinued, and the new government's changing priorities are even very inconsistent with the international mandate to reduce carbon emissions (Pereira & Viola, 2022). Another instance is in Indonesia, where a conflict exists between climate objectives and energy policies (Fünfgeld, 2020). Notwithstanding its commitment to diminish emissions and safeguard forests, Indonesia continues to rely heavily on coal, which accounts for a substantial share of the nation's electrical supply (Adrian et al., 2023). The proliferation of coal mining and coal-fired power facilities has elevated greenhouse gas emissions, undermining the nation's climate obligations (Fünfgeld, 2020; Setyowati & Quist, 2022). Tosun and Peters (2021) divide government policy responses into two types: overreactions, characterized by substantial costs with negligible benefits, and underreactions, which fail to adequately address the complete issue. This occurs when political and economic constraints induce policy inconsistency, causing a nation to implement policies that may be either more ambitious or less robust than those of others.

13.4.1.3 Economic Constraints

A major barrier to political progress in climate diplomacy is the insufficiency of financial resources (Lee, 2021). As it plays a pivotal role in facilitating the transition to a low-carbon and climate-resilient economy, climate finance remains central to the climate discourse(Bhandary et al., 2021; Browne, 2022). Despite global commitments for the $100 billion yearly financial goal set under the United Nations Framework Convention on Climate Change (UNFCCC) and the Paris Agreement, these pledges remain unfulfilled (Bracking & Leffel, 2021; Qi & Qian, 2023). The failure of these financial commitments can be ascribed to several factors such as weak enforcement mechanisms and inconsistent political support (Bracking & Leffel, 2021; Toetzke et al., 2022); inequitable distribution funds (Barrett, 2013); inefficiencies; and inadequate governance capacity (Bhandary et al., 2021; Qi & Qian, 2023). This issue manifests at all levels, including international and national levels, as well as among non-state entities such as corporations and NGOs.

The latest data on climate finance flows have reached an average of 1.3 trillion dollars annually in 2021/2022, marking significant growth compared to previous year (Buchner et al., 2023). However, this progress falls drastically short of the estimated 8–9 trillion dollars annual funding needed by 2030 to meet global climate goals. Geographically, the finance is concentrated in developed economies such as China, the United States, and Europe, while low-income nations and climate-vulnerable regions receive minimal support, with only 3 percent of total funds flowing to least developed countries (LDCs) (ibid). Additionally, a notable transformation in climate funding governance is the growing reliance on non-state actors, especially private sector organizations and households (Bracking & Leffel, 2021). Mechanisms such as green bonds, blended finance initiatives, and public-private partnerships are now crucial to the climate financing sector. Although these methods have effectively utilized supplementary resources, private investors tend to emphasize profit and over-reliance on market-based mechanisms, which reduces the focus on equity and long-term sustainability (Bracking & Leffel, 2021; Kalinowski, 2024).

13.4.1.4 Technological Gaps

Technological innovation is undeniably a significant barrier to climate diplomacy, particularly in underdeveloped nations. The proposed solution for climate actions frequently involves the transfer of low-carbon technologies from developed nations to developing nations, and it requires technology transfer. However, despite numerous UNFCCC procedures facilitating technology transfer, countries in the Global South, such as India, face high capital costs and intellectual property rights (IPRs) that complicate the transfer of hybrid vehicle technology(Ockwell et al., 2008). (Weko & Goldthau, 2022) highlights several barriers to technology transfer, especially in developing countries and least developed countries (LDCs), which encompass in insufficient local competence to assimilate and employ transferred innovations, structural gaps, climate justice disparities, and inadequate financial methods (Weko & Goldthau, 2022).Public-private partnerships and international initiatives frequently neglect the essential local knowledge transfer necessary for the sustainable implementation of these technologies (Bieser, 2024; Fei & Lin, 2017). This leads to dependence on imported technologies without sufficient systems for their maintenance and improvement.

13.4.2 Solutions to Overcome the Barriers: The Interplay of Governance, Leadership, and Public Policy

Climate diplomacy is one of the most critical aspects to address the global challenges of climate change. Despite global efforts, the progress at both the international and national levels has been relatively slow. Various obstacles, including geopolitical tensions, ineffective governance structures, lack of national political will, and economic constraints, limit the collective capacity to successfully tackle this challenge. Addressing these obstacles requires an effective and integrated strategy that utilizes governance, leadership, and public policy to enhance collaboration, stimulate innovation, and ensure equitable execution of climate initiatives. The interaction of these aspects facilitates the navigation of the complexities of climate diplomacy.

Governance frameworks constitute the foundation of climate diplomacy to ensure that global agreements can be translated into effective actions at both the international and national levels. Without a clear system, the global agreements often fall short of their promises. However, as mentioned before, the current governance system often lacks enforcement mechanisms, inclusivity, and transparency (Stankovic et al., 2023). Thus, open reporting and enforcing binding targets are needed to respond to these challenges. Although the international agreement serves to facilitate collective action, international institutions, such as the UNFCCC, can also establish precise standards for monitoring and validating climate projects, thereby preventing any government from disregarding its duties. This is particularly concerning the allocation of responsibilities and objectives between rich nations and vulnerable parties, such as Small Island Developing States (SIDS) and developing countries.

The key factor in achieving effective climate governance relies on domestic policy as it underpins the implementation of international agreements and promotes direct actions. Domestic policies establish the groundwork for incorporating climate objectives into economic and social frameworks, allowing nations to meet their global commitments while managing local issues. As the largest global emitter, China can be used as an example in implementing domestic policy in facing various international climate pressures. China is establishing the Climate Change Department (CCD) and integrating the climate targets into Five-Year Plans, which prioritize economic, environmental, and political priorities (Teng & Wang, 2021). Hence, China's domestic policy has also succeeded in encouraging clean energy investment and increasing capacity in adopting advanced green technology. This makes China one of the leaders in green technology and a larger producer of solar and wind energy technologies (Z. Liu et al., 2021; Zhang et al., 2020).

Leadership, conversely, provides the vision and force necessary to deal with ineffectiveness. Successful leadership in climate diplomacy relies on a nation's capacity to align domestic climate policy with its international commitments, showcasing credibility, influence, and a readiness to foster collective action. Therefore, leadership extends beyond presidents, prime ministers, and other political figures; rather, it may manifest at several levels, including both state and non-state actors. For instance, grassroots leaders frequently contribute essential ideas and inspiration to climate initiatives. Furthermore, the private sector, such as companies and start-ups, can also show leadership in climate diplomacy by innovating in green technology or climate-related solutions. Ultimately, the success of climate diplomacy depends on the interconnected roles of governance, leadership, and public policy.

13.5 Case Studies: Illustrating Political Progress and Barriers (South Korea Green New Deal)

Korea's Green New Deal (GND) Policy provides a compelling case study on how South Korea approaches climate diplomacy. GND policy, initiated in 2020 as a response to the Covid-19 pandemic, aims to transform the economy while addressing social equity and environmental concerns (Han & Lee, 2023). This case study illustrates the synergy of governance, leadership, and public policy practices in tackling climate concerns while fostering economic revitalization.

The Moon Jae-in administration, with the support of the progressive party, adopted the GND as part of the broader K-New Deal (KND) in July 2020 (Han & Lee, 2023). This move came amid the Covid-19 pandemic effect on social and economic conditions as well as the global pressure for climate action (S. H. Lee, 2021). The GND aims to transform the Korean economy from a fossil fuel-dependent, carbon-intensive system to a low-carbon, environmentally friendly economy based on renewable energy. To enhance climate action and achieve a green economy, investments will prioritize green infrastructure (12.1 trillion won from the treasury), renewable energy (24.3 trillion won from the treasury), and the development of the green industry (6.3 trillion won from the treasury) (Government of ROK, 2020).

The governance of Korean GND policy reflects a coordinated and inclusive approach to tackling climate challenges. The policy, led by the president, orchestrates collaboration among government entities, local authorities, and private enterprises. It entails centralized planning for vision formulation (such as the 2050 Carbon Neutral Plan) and localized execution via decentralized implementation. Local governments are assigned the responsibility of adapting the national policy framework to regional circumstances, facilitating decentralized implementation that caters to individual needs, such as localized renewable energy initiatives or green industrial clusters (J.-H. Lee & Woo, 2020). As research done by Han & Lee (2023), compared to previous Korean Green Growth (GG) policy, the GND is more inward-oriented, focused on achieving domestic objectives. Thus, this strategy emphasizes the international agreements, such as Paris Agreement, and establishing measurable targets based on the local context. Additionally, the GND employed financial mechanisms such as tax relief and public funds to promote private-sector involvement (Government of ROK, 2020; Kim et al., 2022; J.-H. Lee & Woo, 2020).

The KGND embodies visionary leadership, seeking to transform Korea from a technical "fast follower" to a "first mover" in the global economy (Government of ROK, 2020). This entails the integration of digital and environmental concerns through the GND with an emphasis on long-term sustainability. The policy that is directly led by the Moon Jae-in administration reflects robust political leadership prioritizing green transformation that is encompassed in the net-zero emissions target and low-carbon-green economy initiatives (Han & Lee, 2023; Kim et al., 2022). At the international level, GND conforms to the international frameworks on climate diplomacy, indicating Korea's commitments to global climate initiatives. This also highlights Korea as the first Asian nation implementing a green recovery strategy after Covid-19 (Han & Lee, 2023). Moreover, Korea's substantial investment in green innovation, including hydrogen energy and electric vehicles, not only diminishes its emissions but also positions the nation as a global center for green technology and climate-related economic diplomacy (Tienhaara et al., 2022).

The Korean GND exemplifies how public policy frameworks can successfully promote climate diplomacy by harmonizing national economic policies with global sustainability goals. The set of policies represents the climate change diplomacy that incorporates economic actions such as opening up green jobs, increasing green investment, assisting Small and Medium Enterprises (SMEs), emission trading schemes, and green technology development, showcasing its role in economic resilience and green industrial growth (J.-H. Lee & Woo, 2020). While the outcomes are still unfolding, the GND demonstrates how public policy frameworks can bridge domestic economic priorities with international climate diplomacy.

However, the Korean GND also faced several barriers to achieving its climate commitments. First, the policy lacks detailed mechanisms for achieving the targets (J.-H. Lee & Woo, 2020). Second, there are budget limitations for establishing energy infrastructures (J.-H. Lee & Woo, 2020). Third, the policies lack details on social equality and international cooperation (Han & Lee, 2023). Additionally, one important benchmark of the Korean GND will be the achievement of energy transition; nevertheless, according to the most recent statistics, Korea continues to depend on fossil fuels in its energy composition, specifically 23.9% coal and 36.5% oil (IEA, 2023). Despite all the barriers, the GND portrays political progress in climate diplomacy, which emphasizes the relations between governance, leadership, and public policy.

13.6 Conclusion

Climate change issues require an integrated and groundbreaking strategy that prioritizes progress in global climate diplomacy. This chapter demonstrates that political progress exceeds formal agreements; it encompasses the combination of governance, leadership, and public policy to overcome barriers and facilitate collaborative action. At the international level, despite all the persistent barriers such as ineffective enforcement mechanisms and capacity gaps between developed and developing countries, climate negotiation has succeeded in moving various countries around the world to jointly deliberate and agree on collective action. At the national level, progress on climate diplomacy is greatly influenced by the policies and leadership of each country in making very context-specific efforts. This is also inseparable from the active role of non-state actors, which also have a pivotal role in advocating the climate agenda.

The challenges to political progress, such as economic constraints, governance deficiencies, and technology disparities, underscore the complicated and interrelated characteristics of the climate issue. Nevertheless, the proposed solutions, including the promotion of polycentric governance frameworks and the enhancement of leadership across several tiers, provide a strategy for overcoming these challenges. The examples discussed in this chapter, such as South Korea's Green New Deal, demonstrate the potential of well-orchestrated political progress to catalyze meaningful change. However, the complexity of climate diplomacy demands continued collaboration, adaptive strategies, and a relentless commitment to addressing the barriers that hinder progress.

Moving forward, the future of climate diplomacy relies on reinforcing the interplay between governance, leadership, and public policy. By establishing more inclusive, transparent, and adaptable mechanisms, the political progress not only meets the urgent demands of climate action but also facilitates equitable and sustainable results. The successes and failures of existing climate initiatives can serve as valuable lessons for future progress in efforts to save the earth from climate change.

Note

1 This work is supported by the National Research Foundation of Korea (RS-2024–00469807) and KEITI (Korea Environmental Industry & Technology Institute) (RS-2023–00221109).

References

Adrian, M., Purnomo, E. P., Enrici, A., & Khairunnisa, T. (2023). Energy transition towards renewable energy in Indonesia. *Heritage and Sustainable Development*, *5*(1), 107–118. https://doi.org/10.37868/hsd.v5i1.108

Agache, I., Aguilera, J., A, C. A., Akdis, M., Barry, M., Bouagnon, A., Chinthrajah, S., Collins, W., Dulitzki, C., Erny, B., Gomez, J., Goshua, A., Jutel, M., Kizer, K., Kline, O., LaBeaud, A. D., Pali-Schöll, I., Perrett, K., Peters, R., … Nadeau, K. (2021). Climate change and global health: A call to more research and more action. *Allergy*, *77*(5), 1389–1407. https://doi.org/10.22541/au.163492618.81492633/v1

Amanatidis, G., & Urbina Trevino, G. (2018). *Climate diplomacy. Workshop proceedings.* European Parliament. https://www.arne-lietz.de/files/2018/03/Climate-diplomacy-proceedings-publication.pdf

Banga, J. (2019). The green bond market: A potential source of climate finance for developing countries. *Journal of Sustainable Finance & Investment*, *9*(1), 17–32. https://doi.org/10.1080/20430795.2018.1498617

Barrett, S. (2013). Local level climate justice? Adaptation finance and vulnerability reduction. *Global Environmental Change*, *23*(6), 1819–1829. https://doi.org/10.1016/j.gloenvcha.2013.07.015

Betsill, M. M., & Corell, E. (Eds.). (2007). *NGO diplomacy: The influence of nongovernmental organizations in international environmental negotiations*. The MIT Press. https://doi.org/10.7551/mitpress/7512.001.0001

Bhandary, R. R., Gallagher, K. S., & Zhang, F. (2021). Climate finance policy in practice: A review of the evidence. *Climate Policy*, *21*(4), 529–545. https://doi.org/10.1080/14693062.2020.1871313

Bieser, J. (2024). *Tackling climate change with AI: Research gaps and needs to align machine learning with greenhouse gas reductions*. https://www.societybyte.swiss/wp-content/uploads/2024/07/Preprint_Bieser_ML-Research-Gaps.pdf

Bowman, M., & Minas, S. (2019). Resilience through interlinkage: The green climate fund and climate finance governance. *Climate Policy*, *19*(3), 342–353. https://doi.org/10.1080/14693062.2018.1513358

Bracking, S., & Leffel, B. (2021). Climate finance governance: Fit for purpose? *WIREs Climate Change*, *12*(4), e709. https://doi.org/10.1002/wcc.709

Browne, K. E. (2022). Rethinking governance in international climate finance: Structural change and alternative approaches. *WIREs Climate Change*, *13*(5), e795. https://doi.org/10.1002/wcc.795

Buchner, B., Naran, B., Padmanabhi, R., Stout, S., Strinati, C., Wignarajah, D., Miao, G., Connolly, J., & Marini, N. (2023, November). *Global landscape of climate finance 2023*. Climate Policy Initiatives.

Bulkeley, H., & Schroeder, H. (2012). Beyond state/non-state divides: Global cities and the governing of climate change. *European Journal of International Relations*, *18*(4), 743–766. https://doi.org/10.1177/1354066111413308

Carlson, D., Robinson, S., Blair, C., & McDonough, M. (2021). China's climate ambition: Revisiting its first nationally determined contribution and centering a just transition to clean energy. *Energy Policy*, *155*, 112350. https://doi.org/10.1016/j.enpol.2021.112350

Çelik, F. B. (2022). The EU's different faces in climate diplomacy: Leadership, interests, and responsibilities. *Journal of European Integration*, *44*(8), 1019–1039. https://doi.org/10.1080/07036337.2022.2068538

Climate Change Act 2008, (2008). https://www.legislation.gov.uk/ukpga/2008/27/data.pdf

Cole, D. H. (2008). Climate vhange and collective action. *Current Legal Problems*, *61*(1), 229–264. https://doi.org/10.1093/clp/61.1.229

Compston, H. (2010). The Politics of climate policy: Strategic options for National Governments. *The Political Quarterly*, *81*(1), 107–115. https://doi.org/10.1111/j.1467-923X.2009.02070.x

Dimitrov, R., Hovi, J., Sprinz, D. F., Sælen, H., & Underdal, A. (2019). Institutional and environmental effectiveness: Will the Paris Agreement work? *WIREs Climate Change*, *10*(4), e583. https://doi.org/10.1002/wcc.583

Dimitrov, R. S. (2016). The Paris Agreement on climate change: Behind closed doors. *Global Environmental Politics*, *16*(3), 1–11. https://doi.org/10.1162/GLEP_a_00361

Dubash, N. K., Hagemann, M., Höhne, N., & Upadhyaya, P. (2013). Developments in national climate change mitigation legislation and strategy. *Climate Policy*, *13*(6), 649–664. https://doi.org/10.1080/14693062.2013.845409

Dye, T. R. (1984). *Understanding Public Policy*. Englewood Cliffs, N.J.

Easton, D. (1953). *The Political System: An Inquiry into the State of Political Science*. Afred A. Knopf.

Fei, R., & Lin, B. (2017). Technology gap and CO2 emission reduction potential by technical efficiency measures: A meta-frontier modeling for the Chinese agricultural sector. *Ecological Indicators*, *73*, 653–661. https://doi.org/10.1016/j.ecolind.2016.10.021

Feola, G., & Nunes, R. (2014). Success and failure of grassroots innovations for addressing climate change: The case of the Transition Movement. *Global Environmental Change*, *24*, 232–250. https://doi.org/10.1016/j.gloenvcha.2013.11.011

Fu, X., & Zhang, J. (2011). Technology transfer, indigenous innovation and leapfrogging in green technology: The solar-PV industry in China and India. *Journal of Chinese Economic and Business Studies*, *9*(4), 329–347. https://doi.org/10.1080/14765284.2011.618590

Fünfgeld, A. (2020). *Coal vs Climate: Indonesia's energy policy contradicts its climate goals*. GIGA German institute of global and area studies – Leibniz-Institut für Globale und Regionale Studien, Institut für Asien-Studien. https://nbn-resolving.org/urn:nbn:de:0168-ssoar-67640-6

Gaora, P. A., Pedrason, R., Herman, E., Irene, I., & Sylvia, S. (2023). Indonesia's climate diplomacy under Joko Widodo: Shaping equitable and sustainable global future. *Nation State: Journal of International Studies*, *6*(1), 34–48. https://doi.org/10.24076/nsjis.v6i1.1026

Giddens, A. (2008). *The politics of climate change: National responses to the challanges of global warming*. Policy Network. https://fcampalans.cat/images/noticias/The_politics_of_climate_change_Anthony_Giddens%282%29.pdf

Government of ROK. (2020). *The Korean New Deal: National strategy for a great transformation*. file:///C:/Users/LENOVO/Downloads/Korean_New_Deal.pdf

Govindaraj, R., & S, S. P. R. (2023). *The influence of international non-governmental organizations and youth movements in propelling global climate norms*. https://doi.org/10.21203/rs.3.rs-2704310/v1

Han, H., & Lee, T. (2023). Varieties of green stimulus policies: Comparative analysis of the green growth and green new deal policies in South Korea. *The Journal of Environment & Development*, *32*(1), 61–82. https://doi.org/10.1177/10704965221139303

Harris, P. G., & Lee, T. (2017). Compliance with climate change agreements: The constraints of consumption. *International Environmental Agreements: Politics, Law and Economics*, *17*(6), 779–794. https://doi.org/10.1007/s10784-017-9365-x

Hsu, A., Moffat, A. S., Weinfurter, A. J., & Schwartz, J. D. (2015). Towards a new climate diplomacy. *Nature Climate Change*, *5*(6), 501–503. https://doi.org/10.1038/nclimate2594

Hulme, M. (2009). *Why we disagree about climate change. Understanding controversy, inaction and opportunity*. University of Cambridge. https://www.researchgate.net/deref/http%3A%2F%2Fideas.repec.org%2Fcgi-bin%2Fhtsearch%3Fq%3DWhy%2BWe%2BDisagree%2BAbout%2BClimate%2BChange?_tp=eyJjb250ZXh0Ijp7ImZpcnN0UGFnZSI6InB1YmxpY2F0aW9uIiwicGFnZSI6InB1YmxpY2F0aW9uIiwicG9zaXRpb24iOiJwYWdlSGVhZGVyIn19

Hulme, M. (2010). The idea of climate change – exploring complexity, plurality and opportunity. *GAIA – Ecological Perspectives for Science and Society*, *19*(3), 171–174. https://doi.org/10.14512/gaia.19.3.4

Huq, S., Kovats, S., Reid, H., & Satterthwaite, D. (2007). Editorial: Reducing risks to cities from disasters and climate change. *Environment and Urbanization*, *19*(1), 3–15. https://doi.org/10.1177/0956247807078058

Hyeonjung, C. (2021). *U.S. climate policy and issues in the Biden era*. The Asan Institute for Policy Studies. https://en.asaninst.org/wp-content/uploads/2021/05/Issue-Brief-Executive-Summary-U.S.-Climate-Policy-and-Issues-in-the-Biden-Era.pdf

IEA. (2023). Korea: Energy mix, What is the role of energy transformation in Korea? International Energy Agency (IEA). https://www.iea.org/countries/korea/energy-mix#what-is-the-role-of-energy-transformation-in-korea

Ibrahim, M. A., & Johansson, M. (2022). Combating climate change – What, where and how to implement adaptive measures in the agriculture sector of Öland, Sweden, keeping in view the constraints of carrying capacities and risk of maladaptation. *Land Use Policy*, *122*, 106358. https://doi.org/10.1016/j.landusepol.2022.106358

Islam, M. (2014). Climate change diplomacy-apparatus for climate change mitigation and adaptation: A reflection in the context of Bangladesh. *British Journal of Environment and Climate Change*, *4*(1), 115–132. https://doi.org/10.9734/BJECC/2014/9174

Jakob, M., Steckel, J. C., Flachsland, C., & Baumstark, L. (2015). Climate finance for developing country mitigation: Blessing or curse? *Climate and Development*, *7*(1), 1–15. https://doi.org/10.1080/17565529.2014.934768

Jordan, A., Huitema, D., Van Asselt, H., & Forster, J. (Eds.). (2018). *Governing climate change: Polycentricity in action?* (1st ed.). Cambridge University Press. https://doi.org/10.1017/9781108284646

Kalinowski, T. (2024). The Green Climate Fund and private sector climate finance in the Global South. *Climate Policy*, *24*(3), 281–296. https://doi.org/10.1080/14693062.2023.2276857

Khatibi, F. S., Dedekorkut-Howes, A., Howes, M., & Torabi, E. (2021). Can public awareness, knowledge and engagement improve climate change adaptation policies? *Discover Sustainability*, *2*(1), 18. https://doi.org/10.1007/s43621-021-00024-z

Kim, S., Lee, M., Yu, I., & Son, J. (2022). Key initiatives for digital transformation, Green New Deal and recovery after Covid-19 within the construction industry in Korea. *Sustainability*, *14*(14), 8726. https://doi.org/10.3390/su14148726

Kinley, R., Cutajar, M. Z., De Boer, Y., & Figueres, C. (2021). Beyond good intentions, to urgent action: Former UNFCCC leaders take stock of thirty years of international climate change negotiations. *Climate Policy*, *21*(5), 593–603. https://doi.org/10.1080/14693062.2020.1860567

Krabbe, O., Linthorst, G., Blok, K., Crijns-Graus, W., Van Vuuren, D. P., Höhne, N., Faria, P., Aden, N., & Pineda, A. C. (2015). Aligning corporate greenhouse-gas emissions targets with climate goals. *Nature Climate Change*, *5*(12), 1057–1060. https://doi.org/10.1038/nclimate2770

Kythreotis, A. P. (2012). Progress in global climate change politics? Reasserting national state territoriality in a 'post-political' world. *Progress in Human Geography*, *36*(4), 457–474. https://doi.org/10.1177/0309132511427961

Leal Filho, W. (2022). Will climate change disrupt the tourism sector? *International Journal of Climate Change Strategies and Management*, *14*(2), 212–217. https://doi.org/10.1108/IJCCSM-08-2021-0088

Lee, E., & Mah, J. S. (2021). Environmental protection and development of technology-intensive industries: The case of new energy vehicle industry in Korea. *Science, Technology and Society*, *26*(3), 413–432. https://doi.org/10.1177/0971721821995589

Lee, J.-H., & Woo, J. (2020). Green New Deal Policy of South Korea: Policy innovation for a sustainability transition. *Sustainability*, *12*(23), 10191. https://doi.org/10.3390/su122310191

Lee, K., & Cheong, I. (2011). Measuring a carbon footprint and environmental practice: The case of Hyundai Motors Co. (HMC). *Industrial Management & Data Systems*, *111*(6), 961–978. https://doi.org/10.1108/02635571111144991

Lee, S. H. (2021). Sustainable Recovery and Economic Transition After the Pandemic in South Korea in Light of the Korean New Deal. In M. Patai & M. Horvath (Eds.), *Age of Eurasia – Future directions of knowledge, technology, money and sustainable geoeconomics. A study volume of the Magyar Nemzeti Bank on the opportunities of Eurasian cooperation* (pp. 277–310). Magyar Nemzeti Bank.

Lee, T. (2021). Financial investment for the development of renewable energy capacity. *Energy & Environment*, 23(6), 1103–1116.

Lee, T., & Morgan, S. (2024). Just energy transition: Concept, components, levels, and actors. *Journal of International Politics*, 29(2), 117–167. http://dx.doi.org/10.18031/jip.2024.12.29.2.117

Liu, H., & Liang, X. (2011). Strategy for promoting low-carbon technology transfer to developing countries: The case of CCS. *Energy Policy*, *39*(6), 3106–3116. https://doi.org/10.1016/j.enpol.2011.02.051

Liu, Z., Deng, Z., He, G., Wang, H., Zhang, X., Lin, J., Qi, Y., & Liang, X. (2021). Challenges and opportunities for carbon neutrality in China. *Nature Reviews Earth & Environment*, *3*(2), 141–155. https://doi.org/10.1038/s43017-021-00244-x

Mayaputri, A., Paramitha, G., & Isnaeni, N. (2024). The island's call for justice: NGO advocacy in the climate litigation case of Pari island in Indonesia. *Technium Social Sciences Journal*, *56*, 282–296. https://doi.org/10.47577/tssj.v56i1.10818

McCubbin, S., Smit, B., & Pearce, T. (2015). Where does climate fit? Vulnerability to climate change in the context of multiple stressors in Funafuti, Tuvalu. *Global Environmental Change*, *30*, 43–55. https://doi.org/10.1016/j.gloenvcha.2014.10.007

Mehrotra, N., & Benjamin, E. O. (2022). Evaluating the enhancement of the nationally determined contributions (NDCs) of developing countries: An international support programme perspective. *Climate Policy*, *22*(6), 728–742. https://doi.org/10.1080/14693062.2022.2071824

Morgan, W., Carter, S. G., & Manoa, F. (2024). Leading from the frontline: A history of Pacific climate diplomacy. *The Journal of Pacific History*, *59*(3), 353–374. https://doi.org/10.1080/00223344.2024.2360093

Nunn, P. (2009). Responding to the challenges of climate change in the Pacific Islands: Management and technological imperatives. *Climate Research*, *40*, 211–231. https://doi.org/10.3354/cr00806

Ockwell, D. G., Watson, J., MacKerron, G., Pal, P., & Yamin, F. (2008). Key policy considerations for facilitating low carbon technology transfer to developing countries. *Energy Policy*, *36*(11), 4104–4115. https://doi.org/10.1016/j.enpol.2008.06.019

Ostrom, E. (2010). Polycentric systems for coping with collective action and global environmental change. *Global Environmental Change*, *20*(4), 550–557. https://doi.org/10.1016/j.gloenvcha.2010.07.004

Pauw, W. P., Klein, R. J. T., Mbeva, K., Dzebo, A., Cassanmagnago, D., & Rudloff, A. (2018). Beyond headline mitigation numbers: We need more transparent and comparable NDCs to achieve the Paris Agreement on climate change. *Climatic Change*, *147*(1–2), 23–29. https://doi.org/10.1007/s10584-017-2122-x

Pereira, J. C., & Viola, E. (2022). Brazilian climate policy (1992–2019): An exercise in strategic diplomatic failure. *Contemporary Politics*, *28*(1), 55–78. https://doi.org/10.1080/13569775.2021.1961394

Peterson, L., Van Asselt, H., Hermwille, L., & Oberthür, S. (2023). What determines climate ambition? Analysing NDC enhancement with a mixed-method design. *Npj Climate Action*, *2*(1), 21. https://doi.org/10.1038/s44168-023-00051-8

Purvis, N., & Stevenson, A. (2010). *Rethinking climate diplomay. New ideas for trasatlatic cooperatio post-Copehagen*. The German Marshall Fund of the United States. http://admin.indiaenvironmentportal.org.in/files/GMFPurvisClimatePaper-web.pdf

Qi, J., & Qian, H. (2023). Climate finance at a crossroads: It is high time to use the global solution for global problems. *Carbon Neutrality*, *2*(1), 31. https://doi.org/10.1007/s43979-023-00071-7

Raiser, K., Kornek, U., Flachsland, C., & Lamb, W. F. (2020). Is the Paris Agreement effective? A systematic map of the evidence. *Environmental Research Letters*, *15*(8), 083006. https://doi.org/10.1088/1748-9326/ab865c

Raustiala, K., & Bridgeman, N. (2007). Nonstate actors in the global climate regime. *SSRN Electronic Journal*. https://doi.org/10.2139/ssrn.1028603

Rechsteiner, R. (2021). German energy transition (Energiewende) and what politicians can learn for environmental and climate policy. *Clean Technologies and Environmental Policy*, *23*(2), 305–342. https://doi.org/10.1007/s10098-020-01939-3

Ryu, S., & Lee, T. (2024). Effects of foreign finance measures on the decoupling of greenhouse gas emissions from economic development: A panel study of 97 developing countries over the period 2000–2018. *Climate & Development*, 16 (7), 588–599. https://doi.org/10.1080/17565529.2023.2273488

Samson, J., Berteaux, D., McGill, B. J., & Humphries, M. M. (2011). Geographic disparities and moral hazards in the predicted impacts of climate change on human populations: Spatially explicit impacts of climate change on human populations. *Global Ecology and Biogeography*, *20*(4), 532–544. https://doi.org/10.1111/j.1466–8238.2010.00632.x

Schroeder, L. D., & Nath, S. (2021). International climate diplomacy, collective action and SIDS. In J. L. Roberts, S. Nath, S. Paul, & Y. N. Madhoo (Eds.), *Shaping the future of small islands* (pp. 325–337). Springer Nature Singapore. https://doi.org/10.1007/978-981-15-4883-3_18

Serrao-Neumann, S., Harman, B., Leitch, A., & Low Choy, D. (2015). Public engagement and climate adaptation: Insights from three local governments in Australia. *Journal of Environmental Planning and Management*, *58*(7), 1196–1216. https://doi.org/10.1080/09640568.2014.920306

Setyowati, A. B., & Quist, J. (2022). Contested transition? Exploring the politics and process of regional energy planning in Indonesia. *Energy Policy*, *165*, 112980. https://doi.org/10.1016/j.enpol.2022.112980

Singh, P. K., & Singh, R. S. (2019). Tackling climate change: A breeder's perspective. In S. Sheraz Mahdi (Ed.), *Climate change and agriculture in India: Impact and adaptation* (pp. 147–162). Springer International Publishing. https://doi.org/10.1007/978-3-319-90086-5_12

Smith, A., Fressoli, M., & Thomas, H. (2014). Grassroots innovation movements: Challenges and contributions. *Journal of Cleaner Production*, *63*, 114–124. https://doi.org/10.1016/j.jclepro.2012.12.025

St. Louis, M. E., & Hess, J. J. (2008). Climate change. *American Journal of Preventive Medicine*, *35*(5), 527–538. https://doi.org/10.1016/j.amepre.2008.08.023

Stankovic, T., Hovi, J., & Skodvin, T. (2023). The Paris Agreement's inherent tension between ambition and compliance. *Humanities and Social Sciences Communications*, *10*(1), 550. https://doi.org/10.1057/s41599-023-02054-6

Stoutenborough, J. W., & Vedlitz, A. (2014). The effect of perceived and assessed knowledge of climate change on public policy concerns: An empirical comparison. *Environmental Science & Policy*, *37*, 23–33. https://doi.org/10.1016/j.envsci.2013.08.002

Stripple, J., & Bulkeley, H. (Eds.). (2013). *Governing the climate: New approaches to rationality, power and politics* (1st ed.). Cambridge University Press. https://doi.org/10.1017/CBO9781107110069

Tänzler, D., & Carius, A. (2013). Beyond international climate negotiations: Climate diplomacy from a foreign policy perspective. In O. C. Ruppel, C. Roschmann, & K. Ruppel-Schlichting (Eds.), *Climate change: International law and global governance* (pp. 259–274). Nomos. https://doi.org/10.5771/9783845242774_259

Taylor, S., & Kumar, L. (2016). Global climate change impacts on Pacific Islands terrestrial biodiversity: A review. *Tropical Conservation Science*, *9*(1), 203–223. https://doi.org/10.1177/194008291600900111

Teng, F., & Wang, P. (2021). The evolution of climate governance in China: Drivers, features, and effectiveness. *Environmental Politics, 30*(sup1), 141–161. https://doi.org/10.1080/09644016.2021.1985221

Tienhaara, K., Yun, S. J., & Gunderson, R. (2022). South Korea's Green New Deal 2.0. In K. Tienhaara & J. Robinson (Eds.), *Routledge handbook on the Green New Deal* (1st ed., pp. 279–298). Routledge. https://doi.org/10.4324/9781003110880-19

Toetzke, M., Stünzi, A., & Egli, F. (2022). Consistent and replicable estimation of bilateral climate finance. *Nature Climate Change, 12*(10), 897–900. https://doi.org/10.1038/s41558-022-01482-7

Tosun, J., & Peters, B. G. (2021). The politics of climate change: Domestic and international responses to a global challenge. *International Political Science Review, 42*(1), 3–15. https://doi.org/10.1177/0192512120975659

UNDP. (2024, January 30). *What is climate change adaptation and why is it crucial?* [Organization Website]. UNDP Climate Promise. https://climatepromise.undp.org/news-and-stories/what-climate-change-adaptation-and-why-it-crucial

United Nations Framework Convention on Climate Change (UNFCCC). (2015). *The Paris Agreement.* United Nations. https://unfccc.int/sites/default/files/resource/parisagreement_publication.pdf

Urwin, K., & Jordan, A. (2008). Does public policy support or undermine climate change adaptation? Exploring policy interplay across different scales of governance. *Global Environmental Change, 18*(1), 180–191. https://doi.org/10.1016/j.gloenvcha.2007.08.002

Vezirgiannidou, S. E. (2015). The UK and emerging countries in the climate regime: Whither leadership? *Global Society, 29*(3), 447–462. https://doi.org/10.1080/13600826.2015.1017554

Wang, D. D., & Sueyoshi, T. (2018). Climate change mitigation targets set by global firms: Overview and implications for renewable energy. *Renewable and Sustainable Energy Reviews, 94*, 386–398. https://doi.org/10.1016/j.rser.2018.06.024

Weko, S., & Goldthau, A. (2022). Bridging the low-carbon technology gap? Assessing energy initiatives for the Global South. *Energy Policy, 169*, 113192. https://doi.org/10.1016/j.enpol.2022.113192

Williams, M. H., & Easton, D. (1954). The political system. *Philosophy and Phenomenological Research, 15*(1), 131. https://doi.org/10.2307/2103757

Woodward, A. (2019). Climate change: Disruption, risk and opportunity. *Global Transitions, 1*, 44–49. https://doi.org/10.1016/j.glt.2019.02.001

Zhang, Y., Sun, J., Yang, Z., & Wang, Y. (2020). Critical success factors of green innovation: Technology, organization and environment readiness. *Journal of Cleaner Production, 264*, 121701. https://doi.org/10.1016/j.jclepro.2020.121701

14

Investment Financing for Climate Change and Economic Sustainability

Toward a Theoretical Framework

Kayira and Darlington Peter Onojaefe

14.1 Introduction

Climate change within the context of investment financing and economic sustainability has gained traction in research and policy with a consideration of practical implications in society. Theunderstanding of investment financing as a necessary theory for climate change mitigation remains an active domain of academic debate. This paper reviews selected theories such as: (1) the theoretical principle of investment financing, (2) modern portfolio theory (MPT), (3) stakeholder theory, and (4) agency theory to understand investment financing suitable for local economic context. While most debate on climate change largely focuses on global players with actors, little is known about the financing requirements that would contribute to the sustainability of the local economy, which mostly suffers from the ravages of climate change and poorly conceived investment financing of projects to boost the local economy (Sayegh, 2018).

Paseda and Okanya (2020) argue that climate change occurs due to human interaction with the environment (though not scientific evidence). Investment financing in infrastructure to support new methods of human interaction with the environment could produce sustained economic outcomes. Though not scientific, Paseda and Okanya provided a new theoretical node to enhance the worldview on climate change. Climate change mitigation through investment financing in infrastructure to promote environmentally friendly economic activities is highlighted as both a sustainable and a progressive element of investment financing in climate change projects. However, the emerging global risk of an increasing multipolar new world order is found to be underscored by existing international agreements like the Paris Agreement (UNFCCC, 2016) and the Sustainable Development Goals (SDGs) (UN, 2015). However, bridging the significant financing gap remains a challenge, with estimates suggesting trillions of dollars are needed annually to meet sustainability targets, particularly in developing economies (World Bank, 2020). While those risks remain potent, a change in the nature of human interaction with the environment would be a small step forward the desired local economic sustainability outcome envisaged in this paper.

14.2 Chapter Objectives and Background

1. To understand the theoretical underpinnings of climate-focused investment financing.
2. To examine financial mechanisms and changes in human behavior toward the environment.
3. To assess the implications of new and emerging technology to support livelihoodresilience.
4. To conclude and recommend a theoretical framework for climate change investment finance.

Theoretical Principle of Investment Financing: A solid theoretical foundation helps contextualize the motivations and mechanisms behind climate-focused investment. This section delves into the main theories that frame sustainable finance. **Modern Portfolio Theory (MPT) and ESG Integration** Modern Portfolio Theory (MPT), developed by Harry Markowitz (1952), argues that investors seek to maximize returns for a given level of risk through diversification. This theory seeks to evaluate financial risk in the context of climate finance to include local environmental and human interaction relativism, environmental, social, and governance (ESG) factors. The inclusion of ESG is primarily aimed at evaluating non-financial risks that might affect long-term economic sustainability relative to specific risks for return on investment. The consideration of ESG for investment finance decisions, like they are now in Pension Funds and Insurance companies, which deny *claims* in times of a natural disaster (Flammer, 2021).

Stakeholder Theory and Corporate Responsibility Stakeholder Theory (Freeman, 1984) suggests that investors should evaluate stakeholders' interests: employees, communities, and the environment in the development and deployment of policy and practice for sound investment financing decisions. The process of risk identification and evaluation provides an opportunity to align investment actions with local economic priorities. CSR initiatives of companies like Unilever (Freeman, 1984).

Agency Theory and Long-term Incentives In 1976, Jensen & Meckling explored agency theory and identified a conflict of interest between stakeholders (principals) and company executives (agents). This conflict of interest is motivated by short-term profit motives that undermine sustainability outcomes to be achieved with climate change-related investment finance. In this context, climate change policy intervention with local economic character may be evaluated and promoted as incentives to would-be investors with a financial reporting credit option and a reward for good corporate citizenship (Jensen & Meckling, 1976). In 2020, Climate Bonds Initiative reported on the Italian energy company (ENEL) that pioneered renewable energy production and distribution for both industrial and domestic consumers. Green bonds enable the company to attract the required investment finance to implement green energy projects. **The Economics of Climate Adaptation** Investing in climate adaptation reduces vulnerability to climate impacts, thereby supporting long-term economic stability. The Adaptation Theory of Change suggests that investments in resilient infrastructure, agricultural systems, and emergency response mechanisms enhance national adaptive capacity, lowering economic losses from climate events (UNFCCC, 2016).

14.3 Methodological Approach

This section provides a detailed overview of the methodological approach used in developing the theoretical framework for investment financing in climate change and economic sustainability. The methodology is grounded in epistemological considerations that frame how knowledge is constructed and validated in this context. Climate change discourse is transient with numerous opinions provided on financing mechanisms, with varied interpretations and motivations. This study integrates constructivist and interpretivist epistemologies to capture the nuanced and evolving dynamics between financial mechanisms, environmental goals, and economic sustainability.

This methodology is designed to support a theoretical inquiry that navigates between abstract concepts (such as sustainability, resilience, and green finance) and empirical insights drawn from industry practices, policy trends, and case studies. **Epistemological Foundations:** the study of knowledge and justification informs how this chapter interprets data and theories in the context of investment financing for climate action. Understanding the epistemological basis of this study is critical, as it shapes the research design, the analysis, and the criteria for proposing and/or opposing views and validating insights. Constructivism posits that knowledge is socially constructed, implying that ideas about climate finance and sustainability are shaped by the perspectives and experiences of various stakeholders (investors, policymakers, corporations, etc.). Interpretivism further allows for an understanding of these stakeholders' subjective interpretations, particularly how they perceive and prioritize different aspects of sustainable finance. **Research design and approach:** this theoretical framework is developed through a qualitative research design involving multiple data sources and analytical methods. The research combines literature review, case studies, and policy analysis. Other sources include databases such as JSTOR, ScienceDirect, and Google Scholar, as well as reports from institutions like the World Bank, United Nations, and Climate Bonds Initiative to construct a cohesive understanding of investment financing for climate action and economic sustainability.

14.4 Literature Review

The urgent need to address climate change and promote economic sustainability has reshaped traditional investment paradigms, leading to the emergence of new financing mechanisms and theoretical frameworks in sustainable finance (Abhayawansa & Adams, 2022). Theoretical perspectives from fields such as environmental economics, finance, and public policy provide insights into how investment financing can be structured to support climate goals while ensuring long-term economic stability (Khan & Munira, 2021). This literature review examines major theoretical frameworks, including sustainable finance, climate finance, stakeholder theory, and agency theory to identify their contributions to an investment framework for climate change and economic sustainability.

Sustainable Finance Theory: Sustainable finance theory posits that financial systems and institutions should incorporate environmental, social, and governance (ESG) considerations to support sustainable development (Shahrour et al., 2023). Traditionally, financial theory has focused on maximizing shareholder wealth without accounting for environmental or social costs (Atsu & Adams, 2021). However, sustainable finance challenges this notion, advocating for a financial system that allocates capital in a way that generates long-term value for all stakeholders. **Environmental, Social, and Governance (ESG) Integration:** ESG integration is a core aspect of sustainable finance theory (Emerson, 2003). The ESG framework evaluates investments based not only on financial returns but also on their environmental impact, social responsibility, and governance practices (Karlsson-Larsson et al., 2025). The theory behind ESG suggests that companies that perform well in these areas are more resilient, especially when faced with environmental or regulatory changes. **Stakeholder Theory:** Stakeholder theory, proposed by Freeman (1984), asserts that corporations should be managed for the benefit of all stakeholders, including employees, customers, communities, and the environment, rather than solely for shareholders. In sustainable finance, stakeholder theory suggests that investments should consider the broader social and environmental impact, balancing the interests of diverse stakeholders. **Agency Theory:** Agency theory, developed by Jensen and Meckling (1976), addresses conflicts of interest between principals (shareholders) and agents (managers). In the context of sustainable finance, agency theory highlights the potential for misalignment between managers' short-term financial goals and investors' long-term sustainability objectives.

Research by Bolton and Kacperczyk (2020) suggests that climate risks, such as regulatory changes or natural disasters, have material impacts on asset values. Theories of risk management, therefore, emphasize the importance of integrating climate risks into financial models, ensuring that investments are resilient to both current and future climate challenges.

14.5 Findings

This chapter presents key findings on the investment financing framework for climate change and economic sustainability. These findings are based on a synthesis of theoretical insights, empirical studies, and case examples in sustainable finance, climate policy, and economic resilience. This section provides insights into how financial mechanisms, market dynamics, stakeholder engagement, and policy frameworks interact to support investments that drive both climate action and economic stability. The findings are organized around four major themes: (1) the effectiveness of financial instruments in supporting climate goals, (2) the role of public-private partnerships in de-risking climate investments, (3) the importance of integrating ESG metrics for economic sustainability, and (4) the challenges and opportunities of regulatory and market-driven frameworks.

Effectiveness of Financial Instruments in Supporting Climate Goals: One of the primary findings is that targeted financial instruments, such as green bonds, climate funds, and blended finance mechanisms, play a crucial role in channeling capital toward climate mitigation and adaptation projects. These instruments have been effective in mobilizing private sector investment while addressing the high upfront costs and risks associated with sustainable projects. **Green Bonds and Environmental Impact:** Green bonds have emerged as a vital instrument in climate finance, enabling governments, corporations, and financial institutions to raise capital specifically for environmental projects (TCFD, 2017). The demand for green bonds has been driven by investors' increasing desire for sustainable assets and alignment with ESG goals. Studies indicate that green bonds can achieve competitive returns while fulfilling climate and environmental goals. For instance, a report by the Climate Bonds Initiative (2021) reveals that the global green bond market surpassed $1 trillion in cumulative issuance, with funds supporting projects in renewable energy, sustainable water management, and energy efficiency.

Risk Mitigation: Green bonds offer a stable, low-risk investment option, attracting a broader range of investors, including institutional investors who are typically more conservative. I**nvestor Demand:** The popularity of green bonds has been driven by the growing interest in ESG investments, indicating that investors are increasingly willing to sacrifice some liquidity for sustainable impact. **Environmental Impact:** Evidence shows that green bonds have financed numerous high-impact projects, contributing to emissions reductions and promoting sustainable infrastructure. However, green bonds face challenges related to transparency and "greenwashing," as not all projects funded by these bonds are fully aligned with rigorous environmental standards. Enhanced certification and reporting frameworks are necessary to ensure the credibility and effectiveness of green bonds. Risk Reduction: By leveraging public funds to reduce risk, blended finance has successfully attracted private investment into renewable energy and climate resilience projects.

14.6 Conclusion and Recommendation

14.6.1 Conclusion

Investment financing frameworks for climate change and economic sustainability are evolving in response to the twin challenges of environmental degradation and the need for economic stability. The research has highlighted the significant role that various financial instruments, such as green bonds, blended finance, and public-private partnerships (PPPs), play in mobilizing capital for climate-resilient projects. These mechanisms have demonstrated effectiveness in attracting private investment, especially when coupled with government support in the form of incentives, subsidies, and regulatory frameworks. However, these efforts face inherent challenges, including regulatory inconsistencies, risks of greenwashing, and gaps in ESG data standardization, which hinder the efficient allocation of capital toward genuinely sustainable projects. Key findings of this chapter suggest that financial markets are increasingly receptive to sustainability considerations, driven by both investor demand and regulatory pressure. ESG metrics are now integral to risk assessment and portfolio management, with evidence showing a positive correlation between high ESG ratings and long-term financial performance. Yet, the challenges of standardizing ESG criteria, ensuring transparency, and preventing greenwashing persist, underscoring the need for robust frameworks and clear reporting standards.

Moreover, blended finance has shown promise as a mechanism to de-risk climate investments in developing markets, where financial and political risks are higher. By combining concessional public capital with private sector investments, blended finance has mobilized substantial resources for high-impact projects, particularly in renewable energy and sustainable infrastructure. However, effective deployment of blended finance requires alignment between public and private interests, clear objectives, and rigorous accountability mechanisms to prevent misallocation of funds. The conclusion also emphasizes the importance of a balanced approach that includes both regulatory frameworks and market-driven mechanisms. While regulatory interventions, such as carbon pricing and mandatory ESG disclosures, establish the necessary incentives and standards, market-driven mechanisms such as voluntary carbon markets and sustainability-linked loans provide flexibility and foster innovation. Both approaches are critical for building a resilient and adaptive investment financing framework capable of addressing the multifaceted nature of climate change and economic sustainability.

14.6.2 Recommendations

To realize the full potential of investment financing for climate change and economic sustainability, several key recommendations are put forth for investors, policymakers, financial institutions, and industry stakeholders. These recommendations aim to strengthen the alignment of finance with climate goals and address the gaps identified in the current financing framework. **Strengthening ESG Integration and Reporting Standards:** Given the critical role of ESG metrics in sustainable finance, it is essential to establish standardized reporting and disclosure frameworks that promote transparency and comparability across sectors and geographies. Standardized ESG metrics would enhance investor confidence and reduce the risks of greenwashing, ultimately leading to more informed investment decisions.

References

Abhayawansa, S., & Adams, C. (2022). Towards a conceptual framework for non-financial reporting inclusive of pandemic and climate risk reporting. *Meditari Accounting Research*, 30(3): 710–738.

Atsu, F., & Adams, S. (2021). Energy consumption, finance and climate change: Does policy uncertainty matter? *Economic Analysis and Policy*, *70*, 490–501.

Bolton, P. and Kacperczyk, M., 2020. Do investors care about carbon risk? *Journal of Financial Economics*, 142(2), 517–549. https://doi.org/10.1016/j.jfineco.2021.05.008

Climate Bonds Initiative (CBI), 2021. *Sustainable debt: Global state of the market 2021*. London: Climate Bonds Initiative. Available at: https://www.climatebonds.net/resources/reports/sustainable-debt-global-state-market-2021 [Accessed 18 August 2025].

Emerson, J. (2003). Blended value proposition: Integrating social and financial returns. *California Management Review*, *45*(4), 35–51.

Flammer, C. (2021). Corporate green bonds. *Journal of Financial Economics, 142*(1), 337–345.

Freeman, R. E. (1984). *Strategic management: A stakeholder approach.* Pitman Publishing.

Jensen, M. C., & Meckling, W. H. (1976). Theory of the firm: Managerial behavior, agency costs, and ownership structure. *Journal of Financial Economics, 3*(4), 305–360.

Karlsson-Larsson,H., Erlandsson, A., Sandberg, J., & Västfjäll, D. (2025). *Mitigating climate change with financial investments: exploring sustainable investment strategies in a novel experimental investment paradigm.* Cambridge University Press.

Khan, R., & Munira, S. (2021). *Climate change adaption as a global public good: implications for financing.* Springer Nature.

Markowitz, H. (1952). Portfolio Selection. *The Journal of Finance, 7*(1), 77–91. https://doi.org/10.2307/2975974

Paseda, O. A., & Okanya, O. (2020). Climate change in the theory of finance. *Journal of Economic and Sustainable Development, 11*(18), 2222–2855.

Sayegh, A. G. (2018). Climate finance: Moral theory and political practice. In J. Tahseen, K. Helwig, & M. Mikulewicz (Eds.), *Routledge handbook of climate justice* (1st ed., pp 153–164). Routledge.

Shahrour, M. H., Arouri, M., & Lemand, R. (2023). On the foundation of firm climate risk exposure. *Review and Accounting and Finance, 22*(5), 620–635.

TCFD. (2017). *Recommendations of the task force on climate-related financial disclosures.*

UN. (2015). *Transforming our world: The 2030 Agenda for sustainable development.* United Nations.

UNFCCC. (2016). *Paris Agreement.* United Nations Framework Convention on Climate Change.

United Nations Framework Convention on Climate Change (UNFCCC), 2019. Annual Report 2019. Bonn: UNFCCC Secretariat. Available at: https://unfccc.int/documents/210328 [Accessed 18 August 2025].

World Bank. (2020). *Transformative climate finance: A new approach for climate finance to achieve low-carbon resilient development in developing countries.* World Bank Group.

15

International Cooperation in Climate Regime

Case Study of Project of Sustainable Life Sustainable Vocational Education

Şefika Şule Erçetin, Kaan Batı, Nilay Neyişci, Şuay Nilhan Açıkalın, Gökhan Dağhan, and Berna Ataberk

15.1 Introduction

The significance of international collaboration in shaping global climate policies is underscored by the development of international frameworks and landmark agreements, including the Kyoto Protocol and the Paris Agreement (Indukuri, 2022; Poddar, 2024). Addressing the global challenge of climate change, which is a quintessential example of the global commons, necessitates concerted efforts through cooperation within international structures (Indukuri, 2022). The United Nations and its associated bodies have been central to advancing climate action, facilitating the adoption of crucial climate agreements such as the Kyoto Protocol and the Paris Agreement (Indukuri, 2022). Similarly, instruments like the Vienna Convention, the Montreal Protocol, and the United Nations Framework Convention on Climate Change (UNFCCC) have played critical roles in shaping global climate policies through collaborative international efforts (Indukuri, 2022). From an economic perspective, current cost-benefit analyses of climate policies often suggest an optimal global policy that falls short of the stringent levels required to meet the internationally agreed-upon 2 degree Celsius target (Cai et al., 2015). These analyses, however, typically fail to incorporate potential environmental tipping points, which could result in abrupt and irreversible impacts on both market and non-market goods and services, potentially necessitating higher carbon taxes than currently proposed (Cai et al., 2015). Furthermore, the United Nations and its institutions, particularly the UNFCCC, have played an instrumental role in promoting climate action and shaping global climate policies through international agreements and treaties (Indukuri, 2022). The Paris Agreement stands out as a pivotal international accord, aiming not only to curb greenhouse gas emissions but also to enhance resilience to climate change and promote sustainable development via international cooperation (Mamalova et al., 2024). In conclusion, international collaboration is integral to the formulation of global climate policies, as evidenced by the central role of international agreements and organizations, the economic implications of climate policies, and the technological innovations needed to address climate change on a global scale. While the provided insights highlight the importance of such collaboration, they do not sufficiently address the specific economic ramifications of global climate policies or the technological advancements supporting these initiatives, indicating a need for further research in these areas. In this book chapter, we present the outputs of a project on sustainable development and sustainable vocational training based on international cooperation.

15.2 Brief Information About Our Project

The Sustainable Life Sustainable Vocational Education Project (Project Code: 2022-1-TR01-KA220-VET000089679) is being implemented by Hacettepe University as part of the Erasmus+ KA220-VET initiative, which focuses on partnerships in vocational education and training. This project is designed to achieve three key objectives: first, to strengthen the competencies and qualifications of both trainers and students in vocational education related to electric bicycles; second, to harmonize vocational education curricula with the latest advancements in electric vehicle technology; and third, to address the demand for mid-level technical professionals in this sector. To accomplish these goals, the project involves the development of modernized training resources for vocational education instructors and the establishment of collaborative networks among universities, vocational training institutions, the EGO General Directorate, and industry stakeholders. A primary focus of the initiative is to equip students with enhanced knowledge and practical skills in the field of electric vehicles, with an emphasis on electric bicycles, while simultaneously integrating innovative, digitally oriented course materials into vocational education programs. The project is being carried out through international collaboration with various partner institutions, including Eu&Pro Centrum Vynálezectví a Praxe, S.R.O. (Czech Republic), ACADEMIA, Izobrazevanje in Druge Storitve D.O.O. (Slovenia), Güvercinlik Şehit Hasan Gülhan Vocational and Technical Anatolian High School (Turkey), EGO General Directorate (Turkey), and Liceul Tehnologic de Transporturi Auto Targoviste (Romania). Through this transnational partnership, the initiative seeks to promote sustainable development and foster innovation within vocational education, particularly in the realm of electric vehicle technologies.

15.2.1 Micro Mobility and Project Justifications

The ongoing increase in urbanization and the resulting growth in the population of people and vehicles has led to significant challenges in ensuring the effective and sustainable mobility of urban populations. The primary sources of these challenges can be attributed to the economic and environmental unsustainability of individual motorized vehicles. Despite the ambitious efforts of urban public transport systems to address these issues, their capacity to provide effective solutions for short-distance transport is limited. Moreover, these systems are unable to adequately respond to the health concerns that may arise from the dense congregation of a large number of people. This has led to a renewed focus on micromobility as a potential solution for urban transportation. The International Transport Forum's (ITF) report, titled "Safe Micro mobility" (ITF, 2020), offers a comprehensive definition of micromobility as: This encompasses vehicles with a mass of no more than 350 kilograms (771 pounds) and a design speed not exceeding 45 kilometer/hour (Santacreu et al., 2020). The definition includes human-powered and electrically assisted vehicles such as bicycles, e-bikes, and e-scooters, as well as skateboards, one-wheeled balance boards, and four-wheeled electric micro-vehicles. Micromobility has emerged as a rapidly expanding mode of urban transportation, particularly favored by younger demographics. These vehicles have increasingly become a staple form of urban mobility, with numerous companies launching micro-vehicles for shared and personal use. The integration of micromobility into urban transport networks holds significant potential in addressing the prevailing challenges in contemporary urban environments. The utilization of micromobility vehicles has the potential to displace the use of personal motorized vehicles, thereby contributing to the reduction of congestion and the enhancement of urban sustainability (Oeschger et al., 2020).

Electric bicycles are recognized as effective tools for sustainable cities and sustainable mobility. However, there are many factors that influence people's decision to purchase electric bicycles. Şimşekoğlu and Klöckner (2019) found that people's intention to purchase e-bikes are positively influenced by their own perceptions of health and mobility, social norms, and familiarity with e-bikes (or bicycles). Consequently, the utilization of electric bicycles within urban environments can be enhanced through the implementation of social awareness initiatives. Moreover, enhancing individual health and environmental consciousness can facilitate the dissemination of micro-mobility (Plazier, 2022). This finding underscores the potential for the popularization of electric vehicles within society. From this perspective, it is imperative to enhance people's purchasing behavior, prioritizing sustainability. Consumer ethics related to sustainability refer to the moral rules, principles, and judgments that enable consumers to make purchasing choices in a sustainable and socially responsible manner, and to use and dispose of products (Muncy & Vitell 1992).

Research on shared micro mobility models has demonstrated that public bicycle sharing enhances urban mobility and curbs motorized vehicle usage (Shaheen et al., 2010). However, global studies on shared bicycle usage reveal that utilization ranges from 0.3 to 2 instances per bicycle per day (Yin et al., 2018). To foster sustainable urban transport, it is imperative to motivate individuals to adopt micro mobility vehicles. Despite the fact that environmentally conscious consumers tend to adopt more environmentally friendly modes of transport and engage in more environmentally friendly behaviors, such as searching for and purchasing ethical products (Chan, 2001), traffic safety remains a primary reason for the low adoption rates of micro mobility despite the practical and environmentally friendly solutions it offers.Basch et al. (2023) reported that there were 4,949 cyclist injuries and 19 fatalities in NYC in 2021. As posited by Basch et al. (2023), approximately one third of cyclists sustaining injuries in bicycle accidents were engaged in occupations such as cargo transportation at the time of the incident. These findings suggest a lack of adequate training and measures in terms of traffic safety. While the dissemination of micro mobility vehicles is advocated for sustainable urban development, the formulation of regulations pertaining to other traffic elements (e.g., other motor vehicles, signaling, and divided roads) should be expedited.

15.3 Project Outcomes

15.3.1 Electric Bicycles Training Program Modules

The document, entitled Electric Bicycles Training Program Modules, serves as a foundational resource for vocational education programs that are aimed at developing expertise in electric bicycle technology. The increasing popularity of electric bicycles as an alternative and sustainable mode of transportation has generated a demand for skilled professionals capable of maintaining and repairing these vehicles. This training program has been meticulously designed to address this emerging need by integrating electric bicycle-related content into vocational education curricula, thereby bridging the gap between theoretical knowledge and practical application. A primary objective of the training program is to furnish students and trainees with a comprehensive understanding of electric bicycle systems, ranging from their historical development to their technical components and regulatory frameworks. The document outlines various educational modules that systematically cover essential topics such as electrical theory, direct current (DC) principles, battery technologies, electric motors, structural features, legal regulations, and periodic maintenance. Each module has been carefully structured to ensure a progressive learning experience that enhances both theoretical knowledge and hands-on technical skills.

A fundamental element of this training program is its interdisciplinary approach, integrating insights from engineering, environmental science, and transportation studies. By examining the historical evolution of electric bicycles, students develop an appreciation for the technological advancements that have shaped modern electric bicycles, from their early prototypes in the late nineteenth century to contemporary high-performance models. The curriculum is further enriched by its exploration of electrical theory and its application in electric bicycles, providing students with the fundamental knowledge required to comprehend power transmission, circuit design, and battery management systems. A significant aspect of the training program is its emphasis on legal and regulatory considerations associated with electric bicycles, which are imperative for ensuring compliance with national and international transportation laws. Students are introduced to the classification of electric bicycles based on motor power and speed, as well as licensing requirements and safety standards that govern their use in different regions. Moreover, the module on periodic maintenance and servicing provides practical guidelines on diagnosing and addressing common mechanical and electrical issues, ensuring that trainees are well-prepared to handle real-world maintenance tasks. By integrating theoretical instruction with hands-on training, this document contributes significantly to the Sustainable Life, Sustainable Vocational Education project. The structured curriculum not only enhances students' employability but also supports the transition toward greener and more sustainable urban transportation solutions. As electric bicycles continue to gain traction worldwide, the development of a skilled workforce through vocational education will be instrumental in facilitating their widespread adoption and long-term sustainability.

15.3.2 Electric Bicycles Training Modules—Student Course

The Electric Bicycles Training Modules—Student Course document is designed as an instructional guide for students enrolled in vocational education programs focusing on electric bicycle technology. The course aims to equip students with the necessary knowledge and skills to understand, operate, and maintain electric bicycles, thereby fostering technical proficiency in a field that is rapidly gaining importance in sustainable transportation. The course is structured into multiple modules, each addressing a specific aspect of electric bicycle systems. The curriculum commences with an exploration of the history and evolution of electric bicycles, thereby providing students with a contextual understanding of their development over time. Subsequent modules introduce fundamental electrical principles, including direct current generation, circuit configurations, and battery management. By incorporating practical applications such as dynamo-based electricity generation tests and motor operation experiments, the course ensures that students gain hands-on experience in working with electric bicycle components. A notable feature of this course is its emphasis on real-world applications. The program incorporates practical exercises that simulate maintenance and troubleshooting scenarios, such as diagnosing motor malfunctions, assessing battery performance, and repairing mechanical failures. These exercises are designed to reinforce theoretical knowledge while developing problem-solving skills that are essential for future employment in the field of electric bicycle maintenance and repair.

Furthermore, the curriculum incorporates a dedicated module on the legal regulations governing electric bicycles. This section familiarizes students with the classification of electric bicycles, licensing requirements, and regional transportation laws, thereby ensuring that students are better prepared to comply with industry standards and ensure the safe operation of electric bicycles in various settings. The Student Course document plays a vital role in the broader e-VETBike initiative, which seeks to integrate electric bicycle education into vocational training. The course offers structured learning experiences, assessment criteria, and performance evaluations, preparing students for careers in sustainable transportation and related industries. As urban mobility trends continue to shift toward eco-friendly alternatives, the skills imparted through this course will be instrumental in supporting the transition to a more sustainable and efficient transportation ecosystem.

15.3.3 Electric Bicycles Training Modules—Teacher Course

The Electric Bicycles Training Modules—Teacher Course document serves as a comprehensive guide for educators involved in vocational training on electric bicycle technology. Its primary objective is to equip teachers with the pedagogical tools and technical expertise required to deliver effective instruction on electric bicycle maintenance, operation, and regulatory compliance. The curriculum outlined in this document is structured to ensure that instructors can effectively convey both theoretical concepts and practical skills to their students. The initial sections provide a detailed overview of the historical development of electric bicycles, emphasizing key technological advancements that have shaped modern electric bicycle systems. Educators are encouraged to integrate this historical perspective into their teaching methodologies, enabling students to appreciate the evolution of electric bicycle technology over time. A significant portion of this document is dedicated to instructional methodologies tailored to vocational education, offering structured lesson plans, laboratory exercises, and assessment strategies to facilitate a hands-on learning experience. Teachers are provided with guidance on conducting practical applications, such as electrical circuit experiments, battery charging tests, and motor operation assessments.

The exercises have been meticulously designed to enhance students' technical competencies and ensure that they acquire the skills necessary for real-world applications. In addition to technical training, the document highlights the importance of legal and regulatory considerations, and educators are provided with resources to teach students about licensing requirements, road safety regulations, and environmental policies related to electric bicycles. This ensures that students not only develop technical expertise but also gain awareness of the broader societal and legal implications of electric bicycle usage. The document's provision of a structured approach to electric bicycle education is of crucial importance in the context of the Sustainable Vocational Education initiative, ensuring consistency in the quality of instruction across different vocational training institutions and supporting the professional development of educators in the field. As demand for electric bicycles continues to rise, the role of well-trained educators will be instrumental in shaping a workforce that is capable of maintaining and advancing sustainable mobility solutions.

15.3.4 Project Contributions to Sustainable Engagement

The project demonstrated considerable potential for expansion across various projects and sectors, owing to its innovative design and flexible methodologies. Its success transcended its initial objectives, positioning it as a model for future endeavors. By focusing on scalability and applicability across multiple fields, the project developed a framework that can be replicated and tailored to address diverse challenges in education, industry, and sustainability. The subsequent section delineates the salient points pertaining to the scalability and cross-sector relevance of the project, elucidating the manner in which its outcomes can be leveraged to foster innovation and progress in diverse domains.

15.3.5 Large-Scale Applications

The e-Learning platform developed for electric bicycles signifies a substantial advancement in the domain of vocational education, offering a multifaceted approach to training and development. The platform incorporates video tutorials, virtual equipment demonstrations, and interactive modules, and its adaptability for extension to other vocational training areas is noteworthy. For instance, its potential for adaptation to fields such as renewable energy, advanced manufacturing, and healthcare technology is evident. The platform's accessibility, including user-friendly interfaces and multilingual support, ensures it caters to diverse target audiences, making it ideal for educational institutions, professional training centers, and corporate learning environments. By utilizing this platform, other initiatives can enhance their training methods, improving engagement, effectiveness, and accessibility.

The project's emphasis on sustainability and the green transition provides a significant framework for application in other industries, with its methods for integrating sustainability into vocational education having the potential for adaptation to sectors such as renewable energy, green architecture, and eco-friendly transportation technologies. For instance, the project's approach to teaching sustainable practices within the context of electric bicycles can be used to train professionals in areas like solar panel installation, energy-efficient building design, or electric vehicle maintenance. The emphasis on sustainability not only addresses current environmental challenges but also equips the workforce for future green economies, rendering the project's results highly relevant and adaptable to other sectors seeking to incorporate sustainability into their training programs.

15.3.6 Applications in Diverse Fields

The training modules developed for electric bicycles provide a valuable model for designing curricula in other emerging technical fields, such as drone technology, autonomous vehicles, and smart grid systems. The systematic approach employed to identify skill gaps and design targeted training programs can be applied to these fields, which, like electric bicycles, require specialized knowledge and hands-on skills. This makes the project's methodologies highly applicable. Furthermore, the project's emphasis on practical, real-world applications ensures that its training modules are not only theoretically robust but also directly aligned with industry needs, facilitating easier adoption and adaptation of its approach by other sectors.

The integration of digital learning tools with sustainability education creates opportunities for cross-disciplinary applications, as evidenced by the following examples. First, the methodologies developed could be utilized in youth training programs to teach both digital literacy and environmental awareness. Second, the project's tools and approaches could be adapted for sports education, incorporating sustainability principles into training programs for athletes and coaches. Thirdly, the project's emphasis on lifelong learning makes it applicable to initiatives focused on upskilling or reskilling professionals across various fields. By fostering cross-disciplinary synergies, the project demonstrates how its outcomes can benefit sectors beyond vocational education.

15.3.7 Benefits of Transnational Collaboration

The involvement of international partners was found to be a pivotal factor in the success of the project, facilitating the exchange of best practices and innovative ideas. This collaborative model has the potential to be replicated in other projects to facilitate knowledge sharing and mutual learning. Insights gained from global partners can assist in the development of training programs that are both culturally relevant and universally applicable, thereby enriching the quality of the outcomes and ensuring their adaptability to diverse contexts and regions. The establishment of networks between vocational institutions, industry leaders, and policymakers has provided a solid foundation for the future, with these partnerships having the potential to expand in order to support larger projects at local, regional, and international levels. For instance, the connections established during this project could be used to launch initiatives focused on emerging technologies or global sustainability challenges. By building on these networks, it is anticipated that other projects will be able to benefit from the expertise, resources, and collaborative spirit that were integral to the project's success.

The project has established the foundations for significant impacts in the domains of education, industry, and sustainability through the implementation of innovative tools and methodologies. Its emphasis on scalability, adaptability, and collaboration positions it as a paradigm for future initiatives that seek to address complex challenges through the application of innovative solutions. The project's outcomes, whether through the expansion of its e-learning platform, the integration of its sustainability practices, or the replication of its collaborative networks, demonstrate its potential to catalyze progress across multiple fields. In achieving these objectives, the project has not only accomplished its immediate goals but also contributed to the creation of a more sustainable, skilled, and interconnected global community.

15.3.8 The Future Role of International Cooperation in Determining Global Climate Policies

The future role of international cooperation in shaping global climate policies is pivotal in addressing the complex and multifaceted challenges posed by climate change. As highlighted by the provided abstracts, the evolution of international climate policy, particularly through significant agreements like the Kyoto Protocol and the Paris Agreement, has had a profound impact on global climate action (Poddar, 2024). The Paris Agreement, in particular, stands as a landmark achievement, representing a consensus among over 190 countries and marking a crucial step toward coordinated global efforts to combat climate change (Stephenson et al., 2019). However, achieving international consensus remains a challenge, with differences in economic status, national interests, and political conflicts often hindering progress (Mamalova et al., 2024). Additionally, there is a growing recognition of the need to enhance emission reduction ambitions to limit global warming to 1.5 degrees Celsius (Mamalova et al., 2024). Emerging economies have increasingly become central to global climate efforts, with variations in environmental outcomes among these economies attributed to differing economic institutions (Benney, 2014). Notably, China's active participation in international climate negotiations, driven by both its domestic needs and aspirations for sustainable development, exemplifies the role of emerging economies in shaping global climate policy (Du, 2014). Future policy development recommendations suggest that developed countries should take the lead in committing to substantial emission reductions, while developing countries should contribute according to their capabilities and developmental stages, supported by funding and technology transfers from developed nations (Du, 2014). Furthermore, the institutionalization of international climate cooperation requires a deep understanding of the values, norms, and principles that drive cooperation, to effectively structure and enforce climate cooperation rules (von Luepke et al., 2024). To enhance climate governance, a comparative framework that explores multilateralism, bilateralism, and transnationalism is necessary, focusing on optimizing the benefits of international collaboration (Liu & Lo, 2021). In conclusion, the future of international cooperation in global climate policies is shaped by the evolution of key agreements, the challenges faced in achieving consensus, the growing influence of emerging economies, and the strategic recommendations for policy development. The insights provided underscore the critical importance of international collaboration in tackling the global climate crisis.

15.4 Discussion and Conclusion

The Sustainable Life Sustainable Vocational Education project is an exemplar of the pivotal function of international collaboration in confronting present-day challenges in climate-conscious vocational training. The initiative underscores the manner in which the integration of sustainability principles into technical education can augment workforce preparedness while concomitantly contributing to overarching environmental objectives. Utilizing an interdisciplinary approach, the project cultivates innovation in electric bicycle technology and establishes a replicable model for analogous endeavors in other sustainable industries. A salient finding of the project is the potential of micro-mobility solutions, particularly electric bicycles, in promoting sustainable urban transportation. This finding is consistent with existing literature emphasizing the economic and environmental advantages of e-bike adoption (Plazier, 2022; Shaheen et al., 2010). However, as research indicates, barriers such as consumer awareness, safety concerns, and regulatory inconsistencies must be systematically addressed to achieve widespread adoption (Basch et al., 2023; Yin et al., 2018).

The project's curriculum development efforts have been shown to contribute to the overcoming of the aforementioned obstacles by equipping both trainers and students with technical and regulatory knowledge, thus ensuring that future professionals are well-prepared to support the green transition in urban mobility (Jones, 2023). Another critical contribution of the project is evidenced by its focus on the alignment of vocational education with industry needs (Schmidt, 2021). The involvement of diverse stakeholders, including universities, training institutions, and industry partners, underscores the significance of aligning educational outcomes with labor market demands (Brown et al., 2022). The project enhances employability by integrating cutting-edge developments in electric bicycle technology into vocational curricula, reinforcing the importance of sustainability-oriented skill sets. Initiatives of this nature are of paramount importance in cultivating a workforce that is equipped to meet the mounting demand for green technologies in the transportation sector and other areas.

Furthermore, the transnational approach adopted by the project offers valuable insights into the potential of international cooperation to enhance the effectiveness of sustainability-driven education. The collaboration among institutions from multiple countries facilitated knowledge exchange and the development of best practices that can be adapted to different regional contexts. This finding lends further support to the broader argument that international frameworks, such as the Paris Agreement, play an essential role in coordinating climate action across borders (Mamalova et al., 2024; Poddar, 2024). The expansion of analogous international partnerships in future educational initiatives has the potential to facilitate innovation in sustainable vocational training by leveraging shared expertise.

In conclusion, the Sustainable Life Sustainable Vocational Education project is an exemplar of the transformative potential of integrating sustainability into vocational education through international cooperation. By addressing key challenges in micro-mobility adoption, bridging educational and industry gaps, and fostering global partnerships, the project contributes significantly to the broader discourse on sustainable urban mobility. Future research should explore the long-term impacts of such initiatives on labor markets and environmental outcomes, further refining strategies for embedding sustainability into vocational training frameworks. The success of the project under discussion highlights the necessity of continued investment in education as a cornerstone of global sustainability efforts.

References

Basch, C. H., Ethan, D., Fera, J., Kollia, B., & Basch, C. E. (2023). Micromobility vehicles, obstructions, and rider safety behaviors in New York City bike lanes. *Journal of Community Health, 48*(3), 522–527.

Benney, T. M. (2014). *Making environmental markets work: The varieties of capitalism in emerging economies.* Routledge.

Brown, P., Souto-Otero, M., & Sadik, S. (2022). Digital transformation and the future of work. In W. Housley et al. (Eds.), The SAGE Handbook of Digital Society. SAGE.

Cai, Y., Judd, K. L., Lenton, T. M., Lontzek, T. S., & Narita, D. (2015). Environmental tipping points significantly affect the cost-benefit assessment of climate policies. *Proceedings of the National Academy of Sciences, 112*(15), 4606–4611.

Chan, R. Y. (2001). Determinants of Chinese consumers' green purchase behavior. *Psychology & Marketing, 18*(4), 389–413.

Du, X. W. (2014). Essence and resolution of international climate negotiation. *Advances in Climate Change Research, 5*(3), 149–152.

Indukuri, S. (2022). International climate change agreements: Setting a global agenda and calling for action. In *Climate change: The social and scientific construct* (pp. 195–213). Springer International Publishing.

International Transport Forum (ITF). (2020). Safe Micromobility. OECD Publishing. https://www.itf-oecd.org/safe-micromobility

Jones, P. (2023). Towards a green and digital transition for European tourism. *Athens Journal of Tourism, 10*(4), 281–294.

Liu, M., & Lo, K. (2021). Pathways to international cooperation on climate governance in China: A comparative analysis. *Journal of Chinese governance, 6*(3), 417–434.

Mamalova, K., Galazova, Z., & Gafarov, E. (2024). Climate agenda: The role of international agreements in combating global warming. *Reliability: Theory & Applications, 19*(81), 669–677.

Oeschger, G., Carroll, P., & Caulfield, B. (2020). Micromobility and public transport integration: The current state of knowledge. *Transportation Research Part D: Transport and Environment, 89*, 102628.

Plazier, P. (2022). E-bikes in rural areas: Current and potential users in the Netherlands. *Transportation, 50*(6), 1–22.

Poddar, A. K. (2024). Climate change and global policy: Traversing the new model of environmental globalization. *The International Journal of Interdisciplinary Global Studies, 19*(2), 125–145.

Santacreu, A., Yannis, G., de Saint Leon, O., & CRIST, P. (2020). *Safe micromobility.* OECD/ITF. 2020. Available online: https://trid.trb.org/view/1696177

Schmidt, T. (2021). Teacher as person: The need for an alternative conceptualisation of the 'good'teacher in Australia's Vocational Education and Training sector. *Journal of Vocational Education & Training, 73*(1), 148–165.

Shaheen, S. A., Guzman, S., & Zhang, H. (2010). Bikesharing in Europe, the Americas, and Asia: Past, present, and future. *Transportation Research Record, 2143*(1), 159–167.

Simsekoglu, O., & Klöckner, C. A. (2019). Factors related to the intention to buy an e-bike: A survey study from Norway. Transportation Research Part F: Traffic Psychology and Behaviour, 60, 573-581. https://doi.org/10.1016/j.trf.2018.11.008

Stephenson, S. R., Oculi, N., Bauer, A., & Carhuayano, S. (2019). Convergence and divergence of UNFCCC nationally determined contributions. *Annals of the American Association of Geographers, 109*(4), 1240–1261.

Vitell, S. J., & Muncy, J. (1992). Consumer ethics: An empirical investigation of factors influencing ethical judgments of the final consumer. *Journal of Business Ethics, 11*, 585–597.

von Luepke, H., Neuhoff, K., & Marchewitz, C. (2024). Bridges over troubled waters: Climate clubs, alliances, and partnerships as safeguards for effective international cooperation? *International Environmental Agreements: Politics, Law and Economics, 24*, 289–308.

Yin, J., Qian, L., & Singhapakdi, A. (2018). Sharing sustainability: How values and ethics matter in consumers' adoption of public bicycle-sharing scheme. *Journal of Business Ethics, 149*, 313–332.

Index

About the Contributors

Şuay Nilhan Açıkalın is currently a faculty member in the Department of International Relations at Hacı Bayram Veli University. Dr. Açıkalın graduated from the Department of International Relations at Bilkent University in 2012. She received her master's degree with the thesis "Leadership in Chaos: Angela Merkel and Euro Crisis" at the Middle East Technical University in 2015. She obtained her PhD degree in 2020 from the same university with her thesis "Political Leadership and Foreign Politics: Recep Tayyip Erdoğan and Angela Merkel." She has been a guest researcher at Lancaster University in the United Kingdom and Humboldt Universität zu Berlin in Germany within the Jean Monnet Scholarship Program. She was a visiting fellow at Mathias Corvinius Collegium in Hungary. Açıkalın works on chaos and complexity theory, leadership, Turkish foreign policy, Turkish-EU relations, Turkey-China relations, women and girls in migration studies, climate change, as well as diplomacy in theory and practice. She is also the author of the book "Leadership Diplomacy in Turkish-German Relations," She is a member of the advisory board of the Turkish Healthy Cities Union and the National Center for the Sea and Maritime Law. She is a founding member of the International Science Association . Dr. Açıkalın is also a member of the Bali Democracy Forum. She has constantly given interviews to national and international media outlets such as TRTWORLD, Al Jazeera, and so on.

Amit Upadhyay is an associate professor at O.P. Jindal Global University and holds an LLM in European and International Law from Freie Universität, Berlin, Germany. His research interests include Constitutional Law and Human Rights Law. He has been awarded several scholarships and fellowships to participate in international academic conferences and programs by organizations such as the Robert Bosch Foundation, German Academic Exchange Service (DAAD), BMZ, Konrad Adenauer Foundation, and the International Institute of Human Rights, Strasbourg. His recent publications include a co-authored article titled "A case for recognition of caste as a sui generis category in international law," published in the Hungarian Journal of Legal Studies in 2024.

Abhinav Mehrotra is an Associate Professor, Jindal Global Law School, O.P. Jindal Global University and Assistant Director, Centre for the Study of United Nations and holds an LLM in International Human Rights Law from the University of Leeds. His research interests include Public International Law and Human Rights Law. He has been published in various reputable journals internationally and has attended various international conferences on Human Rights. His recent publications include a co-authored article titled "India's Approach towards Adjudication before the International Court of Justice and International Criminal Court: in Search of Uniformity," published in JANUS. NET, e-journal of International Relations, run by the Observatory for External Relations, University of Lisbon, Portugal, in 2024.

Zeynep Olgun completed her undergraduate education at Ondokuz Mayıs University, Sinop Faculty of Education, Department of Preschool Teaching, her graduate education at Gazi University, Institute of Educational Sciences, and her doctorate at Hacettepe University, Department of Educational Administration in 2024. She has national and international publications in the fields of educational management, educational leadership, higher education, internationalization, lifelong education, preschool education, and scale development. She works as a specialist teacher at the Ministry of National Education.

Brígida Brito is Associate Professor, Deputy Director, and Head of Quality at the Department of International Relations, Universidade Autónoma de Lisboa (Portugal), Deputy Director of OBSERVARE, and Editor-in-Chief of *JANUS. NET*, e-journal of international relations. She is an integrated researcher at the Portuguese Institute of International Relations of the New University of Lisbon (IPRI-NOVA). She is a sociologist, has a PhD in African Studies and a master's degree in the same scientific field. She completed a post-doctorate in socio-environmental studies in Africa. She participates in and organizes scientific meetings and has numerous articles and chapters in reputable publications.

Vesselin Popovski is the Founding Executive Director of the Centre for the Study of the United Nations, Vice Dean (Institution Building) of the Jindal Global Law School, Senior Advisor of the International Institute for Higher Education and Research, and Advisory Council Member of the Office for Interdisciplinary Studies at O.P. Jindal Global University. In 2004–14 he was the Senior Academic Officer and Head of "Peace and Governance" Programme at UN University, Tokyo. Prior to that, he was Co-Director of the EU project "Legal Protection of Individual Rights in Russia" and an Assistant Professor at

Exeter University, United Kingdom. He has a PhD from King's College London, an M.Sc. from London School of Economics, and BA/MA from the Moscow Institute of International Affairs. He has written or edited over twenty-five books and numerous articles in academic journals.

Taedong Lee is the Underwood Distinguished Professor in the Department of Political Science and International Relations at Yonsei University, Seoul, and Director of the Environment, Energy, and Human Resource Development Center. He leads pioneering initiatives such as the Climate Adaptation Living Lab Research & Development and Project Y: Transformative Climate Change Research & Education. His research specializes in environmental governance, examining it through the perspectives of international relations and public policy.

Dian Wahyu Utami is a PhD student at Yonsei University, working at the Environmental, Energy, and Urban Politics Lab. Their research focuses on climate diplomacy, sustainable development, and energy politics. While currently building a foundation in these fields, Dian's work reflects a deep interest in contributing to the global dialogue on environmental governance and climate policy.

Şefika Şule Erçetin is the Dean of the Faculty of Education at Hacettepe University, Türkiye. She is currently the Dean of the Education Faculty at Hacettepe University,Ankara, Türkiye. In 1991, she completed her PhD from the Division of Educational Administration, Supervision, Planning, and Economics at Hacettepe University. Since 2003, she has been working as a professor. She worked as a visiting scholar at Texas A&M, United States of America, and Lancaster University, United Kingdom. She has worked on many national and international projects as a member of the project team and as the project coordinator on the education of refugees, women, and vocational training. She was the Founder of the International Symposium on Chaos, Complexity and Leadership. She is one of the rare academics in Turkey whose interests are in "chaos, complexity, quantum leadership" since the 1990s. Additionally, she is the author and editor of books on complexity theory from Springer, World Scientific, and Apple Academic Press related to comparative education policies, research methods, organizational intelligence, stupidity and wisdom, leadership, migration, fuzzy logic, children at risk, women and peace studies, social media, plasma leadership, quantum leadership, and women leadership models. Furthermore, she is the President, Board Member, and Founding Member of the International Science Association in Türkiye (ISCASS). She is the Editor-in-Chief of various national and international journals that focus on leadership and education.

Nurşah Ataş is a research assistant in the Department of Foreign Language Education at Hacettepe University, Faculty of Education, since 2023. She is currently pursuing her PhD in the same department. She has actively contributed to the organization of various national and international projects and conferences. Her research interests primarily include digital learner identity, informal digital learning of English, and artificial intelligence-mediated language teaching and learning.

H. Burç Aka graduated from the Department of Political Science and Public Administration at Bilkent University. He completed his master's degree at Kadir Has University and earned his doctorate from the European Union Institute at Marmara University with his dissertation titled "Security Dimensions of European Integration." Since 2022, he has been serving as a professor of international relations at Tekirdağ Namık Kemal University. His research focuses on international security, the analysis of Turkish foreign policy. He has published academic articles and book chapters in national and international outlets.

Sezin İba Gürsoy,, is an associate professor at Kırklareli University, Department of International Relations. She completed her doctorate degree at Istanbul University in 2013. In her thesis, she examined food security within the scope of environmental security, at the scale of Turkey and the European Union. Dr. Gürsoy worked as a visiting researcher at the University of Central Florida in 2015. Her research areas mostly include water and food security, transboundary environmental problems, climate change, and European Union-Turkey relations. She has articles published in journals such as Water International, and the Journal of Environmental Studies and Sciences. She also conducted a scientific research project to increase environmental awareness among university students. She has regular articles in T24 Sunday supplements on international politics, especially the environmental agenda.

Rajat Shandilya is a Full-Time Doctoral Research Scholar (Law) at Symbiosis Law School, Noida, Symbiosis International (Deemed to be) University, Pune, India. Rajat also serves as a freelance Research Assistant at the Centre for the

Study of the United Nations (CSUN), Jindal Global Law School, O.P. Jindal Global University, Sonipat, Haryana, India. His research interests include Criminal Law, Criminal Justice, and Environmental Law, among others.

Betül Uygur was born in Izmir in 1990. In 2013, she graduated from the Faculty of French Translation and Interpreting at Hacettepe University, Ankara. She spent one semester in Mons, Belgium, within the Erasmus Exchange Program and studied at the Faculty of Translation and Interpreting at the University of Mons. In 2021, she received her Master's degree from Ankara University, Faculty of Political Science, Department of African Studies, with the thesis "The Question of the Wall in Western Sahara." She is a PhD candidate at the International Relations Department of Hacı Bayram Veli University. She is fluent in English and French (advanced level), Italian, and Spanish (beginner level). She has been working in the media department of the Presidency of the Republic of Türkiye since 2014.

Mithat EKİCİ graduated from Anatolian Meteorological High School in 1990. In the same year, he started working as a weather forecaster at the Erzurum Meteorology Regional Directorate. In 1993, he began his undergraduate education at Ege University—Faculty of Agriculture—Department of Field Crops and was appointed to the Izmir Meteorology Regional Directorate. He completed his undergraduate education in 1998. He was appointed to the General Directorate of Meteorology in 2000. He worked in the ozone layer research unit as an engineer for about ten years. He continues his climate and climate change studies along with his ozone layer studies. In 2011, he received his MA in Public Administration and Political Science from TODAIE-Türkiye and the Middle East Public Administration Institute. In 2022, he graduated from Ankara Hacı Bayram Veli University—Faculty of Economics and Administrative Sciences—Public Administration and Political Science department with his thesis titled "Turkey's Climate Change Policy as a Public Policy." He currently works as the Climate and Climate Change Branch Manager at the Turkish State Meteorological Service. He is married and has two children.

Sinem Sinangil is a PhD student in Social Work at Hacettepe University, specializing in climate change adaptation, migration, and social policy. She holds a Master's degree in Political Science from Hacettepe University and has conducted academic research at Utrecht University in the Netherlands. She also gained valuable policy experience in Brussels, contributing to international discussions on climate governance and sustainability. With a strong background in speechwriting and policy analysis, she serves as a member of the Climate Council Commission on Climate Change Adaptation, where she actively works on shaping environmental policies and governance frameworks. Her research focuses on climate resilience, social justice, and the impact of environmental change on vulnerable communities. Based in Ankara, she continues to engage in academic research, international collaborations, and policy advisory roles, advocating for climate justice and sustainable development.

Paulo Magalhães is a jurist and researcher at the CIJ—Centre for Interdisciplinary Research on Justice of the Faculty of Law of the University of Porto. He graduated from the Catholic University of Porto, completed a postgraduate at the University of Coimbra, earned his PhD in Human Ecology at the New University of Lisbon, and completed his postdoctoral at the Faculty of Law of the University of Porto with work on the legal status of climate. He is the author of "The Earth Condominium—From the Climate Change to a New Juridic Conception of the Planet," 2007, and editor of "SOS—Safe Operating Space Treaty: A new approach to manage our use of the Earth System," 2016. He is the founder and Executive Director of Common Home of Humanity (Casa Comum da Humanidade) and coordinator of the Task Force for the recognition of climate as common heritage of humankind. He is a Counsellor of CNADS—National Council for Sustainable Development, and a member of the Scientific Committee of the Portuguese Council for Health and Environment. He received the Green Vision Inspiration Award in 2022, the Gold Medal of Merit from the city of Porto, and the Inspiring Portugal Award in Social Economy in 2022.

Mohd Imran is a lecturer at the Faculty of Shariah and Law, Villa College, Maldives, where he specializes in public international law, European Union law, and international relations. His research and teaching focus on critical issues such as climate change, foreign investment, and geopolitics of small island nations. He is currently involved in a UNDP-funded project advocating for a regional agreement on environmental protection, focusing on climate change, marine protection, and biodiversity. Mohd's work reflects his commitment to addressing global challenges through legal frameworks and international cooperation.

Aishath Maahy Shafeeu is a researcher at the Faculty of Shariah and Law, Villa College, Maldives. Previously, she has worked at the Human Rights Commission of the Maldives (HRCM), where she honed her research and analytical skills. Passionate about global issues, Aishath is particularly interested in geopolitics, climate change, and issues concerning small island nations.

Shadi Ibrahim Al-Banna is a writer, educator, and researcher. He holds a PhD in Educational Administration from Hacettepe University and completed both a BA in English and an MA in Educational Administration from the Islamic University of Gaza. Additionally, he earned diplomas in Turkish and Hebrew. Throughout his career, Shadi has served as an educational supervisor, content manager, and international trainer. He is also the founder of the Palestinian Students Forum in Turkey and remains an active contributor to academic research and conferences.

Kaan Batı his bachelor's degree in science education from Hacettepe University in 2005, his master's degree in science education in 2010, and his PhD degree in science education in 2014 from Hacettepe University. He prepared his master's thesis on scientific process skills and problem-solving, and his doctoral thesis on modeling, the nature of science, and critical thinking. In the 2018–19 academic year, he completed his postdoctoral studies at The Ohio State University (USA) within the scope of the TÜBİTAK 2219A project.

www.ingramcontent.com/pod-product-compliance
Ingram Content Group UK Ltd.
Pitfield, Milton Keynes, MK11 3LW, UK
UKHW020217290926
15470UKWH00001B/249

9 781666 978230